Gas Turbine Engineering

Gas Turbine Engineering

Edited by
Zachary Morgan

🖩 Larsen & Keller
www.larsen-keller.com

Gas Turbine Engineering
Edited by Zachary Morgan
ISBN: 978-1-63549-131-9 (Hardback)

© 2017 Larsen & Keller

🖥 Larsen & Keller

Published by Larsen and Keller Education,
5 Penn Plaza,
19th Floor,
New York, NY 10001, USA

Cataloging-in-Publication Data

Gas turbine engineering / edited by Zachary Morgan.
      p. cm.
Includes bibliographical references and index.
ISBN 978-1-63549-131-9
1. Gas-turbines. 2. Turbines. I. Morgan, Zachary.
TJ778 .G37 2017
621.433--dc23

The publisher's policy is to use permanent paper from mills that operate a sustainable forestry policy. Furthermore, the publisher ensures that the text paper and cover boards used have met acceptable environmental accreditation standards.

Printed and bound in the United States of America.

For more information regarding Larsen and Keller Education and its products, please visit the publisher's website www. larsen-keller.com

Table of Contents

Preface

The science of gas turbines is an ancient one and it has been evolving rapidly since its discovery. The rise of new technology has positively affected the development process of this technology. Gas turbines are used to power ships, tanks, aircrafts, trains and even electric generators. The book aims to shed light on some of the unexplored aspects of gas turbine technology. It is a valuable compilation of topics, ranging from the basic to the most complex theories and principles in the field of gas turbines. The topics covered in this extensive book deal with the core subjects of this field. Different approaches, evaluations and methodologies on gas turbine engineering have been included in this compilation. Coherent flow of topics, student-friendly language and extensive use of examples make this textbook an invaluable source of knowledge. It will serve as a reference to a broad spectrum of readers.

A detailed account of the significant topics covered in this book is provided below:

Chapter 1- Gas turbine is a type of internal combustion engine whose operation is similar to that of a steam power plant except that it uses air instead of water. They are used to power aircraft, trains, ships, tanks etc. This chapter provides the reader with an introductory account on turbines, steam turbines, gas turbines, stator, turbine map and their applications.

Chapter 2- The major components of a gas turbine are a gas compressor and a combustion chamber. The gas compressor pumps in atmospheric air and brings it to high pressure while fuel is sprayed and ignited in the combustion chamber to produce energy. The chapter details the whole process that converts fuel into energy with detailed step by step information about the functioning of a gas turbine. This chapter is an overview of the subject matter incorporating all the major aspects of gas turbines.

Chapter 3- Gas turbines are classified by the fuel they use and the thrust they produce. Different types of gas turbines seek to maximize the form of energy utilized. This chapter focuses on various types of gas turbines like turboprop, jet engine, auxiliary power unit, turboshaft, closed-cycle gas turbine, radial turbine, turbopump etc. providing valuable information about their technological aspects, uses, application and their types.

Chapter 4-The Brayton cycle, a thermodynamic cycle can be used to explain the workings of a heat engine. The chapter elucidates on the salient features of the Brayton cycle and its two types. The chapter also explores the concepts of an internal combustion engine to aid in the better understanding of gas turbines.

Chapter 5- This chapter discusses the methods of gas turbines in a critical manner by providing key analysis of the subject matter. Topics explored in this chapter include air-start system, axial compressor, turbine inlet air cooling and overall pressure ratio. This chapter discusses the methods of gas turbines in a critical manner providing key analysis to the subject matter.

Chapter 6- Due to the extensive applications of gas turbines in the field of power generation, they are also employed in related fields. This chapter provides the reader with a thorough understanding of the applications of gas turbines in the fields of marine propulsion, trains, electric locomotives, propfans, aircrafts etc.

It gives me an immense pleasure to thank our entire team for their efforts. Finally in the end, I would like to thank my family and colleagues who have been a great source of inspiration and support.

Editor

Introduction to Gas Turbines

Gas turbine is a type of internal combustion engine whose operation is similar to that of a steam power plant except that it uses air instead of water. They are used to power aircraft, trains, ships, tanks etc. This chapter provides the reader with an introductory account on turbines, steam turbines, gas turbines, stator, turbine map and their applications.

Gas Turbine

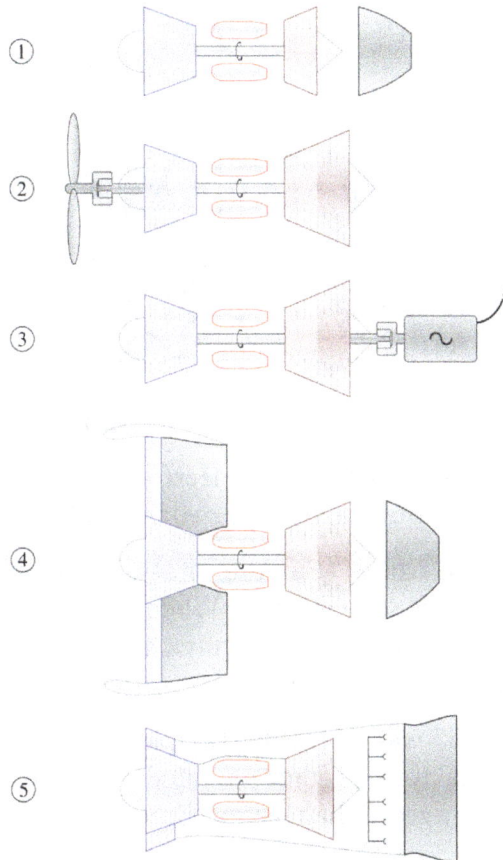

Examples of gas turbine configurations: (1) turbojet, (2) turboprop, (3) turboshaft (electric generator), (4) high-bypass turbofan, (5) low-bypass afterburning turbofan

A gas turbine, also called a combustion turbine, is a type of internal combustion engine. It has an upstream rotating compressor coupled to a downstream turbine, and a combustion chamber in between.

The basic operation of the gas turbine is similar to that of the steam power plant except that air is used instead of water. Fresh atmospheric air flows through a compressor that brings it to higher pressure. Energy is then added by spraying fuel into the air and igniting it so the combustion generates a high-temperature flow. This high-temperature high-pressure gas enters a turbine, where it expands down to the exhaust pressure, producing a shaft work output in the process. The turbine shaft work is used to drive the compressor and other devices such as an electric generator that may be coupled to the shaft. The energy that is not used for shaft work comes out in the exhaust gases, so these have either a high temperature or a high velocity. The purpose of the gas turbine determines the design so that the most desirable energy form is maximized. Gas turbines are used to power aircraft, trains, ships, electrical generators, and tanks.

History

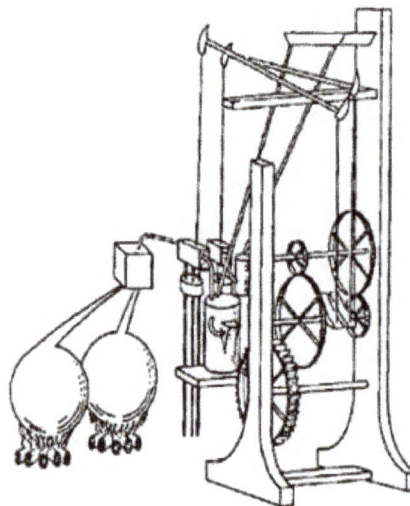

Sketch of John Barber's gas turbine, from his patent

- 50: Hero's Engine (*aeolipile*) — Apparently, Hero's steam engine was taken to be no more than a toy, and thus its full potential not realized for centuries.

- 1000: The "Trotting Horse Lamp" (Chinese: 走马灯) was used by the Chinese at lantern fairs as early as the Northern Song dynasty. When the lamp is lit, the heated airflow rises and drives an impeller with horse-riding figures attached on it, whose shadows are then projected onto the outer screen of the lantern.

- 1500: The "Chimney Jack" was drawn by Leonardo da Vinci: Hot air from a fire rises through a single-stage axial turbine rotor mounted in the exhaust duct of the fireplace and turning the roasting spit by gear/ chain connection.

- 1629: Jets of steam rotated an impulse turbine that then drove a working stamping mill by means of a bevel gear, developed by Giovanni Branca.

- 1678: Ferdinand Verbiest built a model carriage relying on a steam jet for power.

- 1791: A patent was given to John Barber, an Englishman, for the first true gas turbine. His invention had most of the elements present in the modern day gas turbines. The turbine was designed to power a horseless carriage.

- 1861: British patent no. 1633 was granted to Marc Antoine Francois Mennons for a "Caloric engine". The patent shows that it was a gas turbine and the drawings show it applied to a locomotive. Also named in the patent was Nicolas de Telescheff (otherwise Nicholas A. Teleshov), a Russian aviation pioneer.

- 1872: A gas turbine engine was designed by Franz Stolze, but the engine never ran under its own power.

- 1894: Sir Charles Parsons patented the idea of propelling a ship with a steam turbine, and built a demonstration vessel, the *Turbinia*, easily the fastest vessel afloat at the time. This principle of propulsion is still of some use.

- 1895: Three 4-ton 100 kW Parsons radial flow generators were installed in Cambridge Power Station, and used to power the first electric street lighting scheme in the city.

- 1899: Charles Gordon Curtis patented the first gas turbine engine in the USA ("Apparatus for generating mechanical power", Patent No. US635,919).

- 1900: Sanford Alexander Moss submitted a thesis on gas turbines. In 1903, Moss became an engineer for General Electric's Steam Turbine Department in Lynn, Massachusetts. While there, he applied some of his concepts in the development of the turbo-supercharger. His design used a small turbine wheel, driven by exhaust gases, to turn a supercharger.

- 1903: A Norwegian, Ægidius Elling, was able to build the first gas turbine that was able to produce more power than needed to run its own components, which was considered an achievement in a time when knowledge about aerodynamics was limited. Using rotary compressors and turbines it produced 11 hp (massive for those days).

- 1906: The Armengaud-Lemale turbine engine in France with water-cooled combustion chamber.

- 1910: Holzwarth impulse turbine (pulse combustion) achieved 150 kilowatts.

- 1913: Nikola Tesla patents the Tesla turbine based on the boundary layer effect.

- 1920s The practical theory of gas flow through passages was developed into the more formal (and applicable to turbines) theory of gas flow past airfoils by A. A. Griffith resulting in the publishing in 1926 of *An Aerodynamic Theory of Turbine Design*. Working test-bed designs of axial turbines suitable for driving a propellor were developed by the Royal Aeronautical Establishment proving the efficiency of aerodynamic shaping of the blades in 1929.

- 1930: Having found no interest from the RAF for his idea, Frank Whittle patented the design for a centrifugal gas turbine for jet propulsion. The first successful use of his engine was in April 1937.

- 1932: BBC Brown, Boveri & Cie of Switzerland starts selling axial compressor and turbine turbosets as part of the turbocharged steam generating Velox boiler. Following the gas turbine principle, the steam evaporation tubes are arranged within the gas turbine combustion chamber; the first Velox plant was erected in Mondeville, France.

- 1934: Raúl Pateras de Pescara patented the free-piston engine as a gas generator for gas turbines.

- 1936: Hans von Ohain and Max Hahn in Germany were developing their own patented engine design.

- 1936 Whittle with others backed by investment forms Power Jets Ltd

- 1937 The first Power Jets engine runs, and impresses Henry Tizard such that he secures government funding for its further development.

- 1939: First 4 MW utility power generation gas turbine from BBC Brown, Boveri & Cie. for an emergency power station in Neuchâtel, Switzerland.

- 1946 National Gas Turbine Establishment formed from Power Jets and the RAE turbine division bring together Whittle and Hayne Constant's work. In Beznau, Switzerland the first commercial reheated/recuperated unit generating 27 MW was commissioned.

- 1963 Pratt and Whitney introduce the GG4/FT4 which is the first commercial aeroderivative gas turbine.

- 2011 Mitsubishi Heavy Industries tests the first >60% efficiency gas turbine (the M501J) at its Takasago works.

Theory of Operation

In an ideal gas turbine, gases undergo three thermodynamic processes: an isentropic compression, an isobaric (constant pressure) combustion and an isentropic expansion. Together, these make up the Brayton cycle.

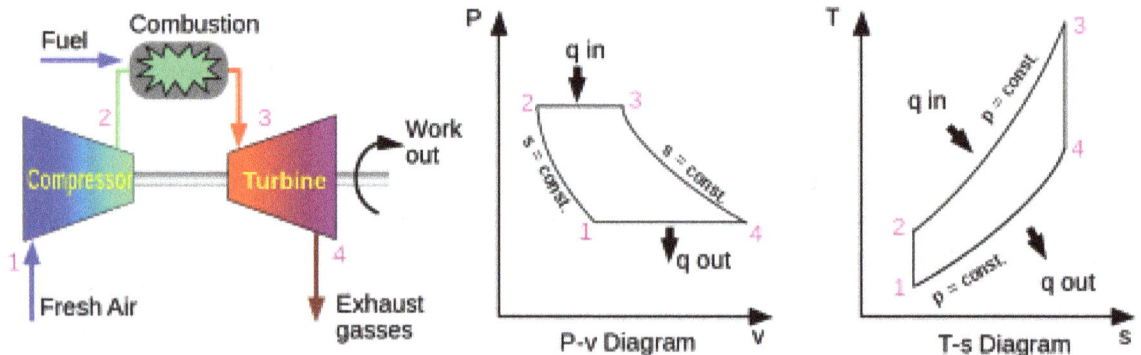

Brayton cycle

In a real gas turbine, mechanical energy is changed irreversibly (due to internal friction and turbulence) into pressure and thermal energy when the gas is compressed (in either a centrifugal or axial compressor). Heat is added in the combustion chamber and the specific volume of the gas increases, accompanied by a slight loss in pressure. During expansion through the stator and rotor passages in the turbine, irreversible energy transformation once again occurs.

If the engine has a power turbine added to drive an industrial generator or a helicopter rotor, the exit pressure will be as close to the entry pressure as possible with only enough energy left to overcome the pressure losses in the exhaust ducting and expel the exhaust. For a turboprop engine there will be a particular balance between propeller power and jet thrust which gives the most economical operation. In a jet engine only enough pressure and energy is extracted from the flow to drive the compressor and other components. The remaining high pressure gases are accelerated to provide a jet to propel an aircraft.

The smaller the engine, the higher the rotation rate of the shaft(s) must be to attain the required blade tip speed. Blade-tip speed determines the maximum pressure ratios that can be obtained by the turbine and the compressor. This, in turn, limits the maximum power and efficiency that can be obtained by the engine. In order for tip speed to remain constant, if the diameter of a rotor is reduced by half, the rotational speed must double. For example, large jet engines operate around 10,000 rpm, while micro turbines spin as fast as 500,000 rpm.

Mechanically, gas turbines can be considerably less complex than internal combustion piston engines. Simple turbines might have one main moving part, the compressor/shaft/turbine rotor assembly, with other moving parts in the fuel system. However, the precision manufacture required for components and the temperature resistant alloys necessary for high efficiency often make the construction of a simple gas turbine more complicated than a piston engine.

More advanced gas turbines (such as those found in modern jet engines) may have 2 or 3 shafts (spools), hundreds of compressor and turbine blades, movable stator blades, and extensive external tubing for fuel, oil and air systems.

Thrust bearings and journal bearings are a critical part of design. Traditionally, they have been hydrodynamic oil bearings, or oil-cooled ball bearings. These bearings are being surpassed by foil bearings, which have been successfully used in micro turbines and auxiliary power units.

Creep

A major challenge facing turbine design is reducing the creep that is induced by the high temperatures. Because of the stresses of operation, turbine materials become damaged through these mechanisms. As temperatures are increased in an effort to improve turbine efficiency, creep becomes more significant. To limit creep, thermal coatings and superalloys with solid-solution strengthening and grain boundary strengthening are used in blade designs. Protective coatings are used in to reduce the thermal damage and to limit oxidation. These coatings are often stabilized zirconium dioxide-based ceramics. Using a thermal protective coating limits the temperature exposure of the nickel superalloy. This reduces the creep mechanisms experienced in the blade. Oxidation coatings limit efficiency losses caused by a buildup on the outside of the blades, which is especially important in the high-temperature environment. The nickel-based blades are alloyed with aluminum and titanium to improve strength and creep resistance. The microstructure of these alloys is composed of different regions of composition. A uniform dispersion of the gamma-prime phase – a combination of nickel, aluminum, and titanium – promotes the strength and creep resistance of the blade due to the microstructure. Refractory elements such as rhenium and ruthenium can be added to the alloy to improve creep strength. The addition of these elements reduces the diffusion of the gamma prime phase, thus preserving the fatigue resistance, strength, and creep resistance.

Types

Jet engines

typical axial-flow gas turbine turbojet, the J85, sectioned for display. Flow is left to right, multistage compressor on left, combustion chambers center, two-stage turbine on right

Airbreathing jet engines are gas turbines optimized to produce thrust from the exhaust gases, or from ducted fans connected to the gas turbines. Jet engines that produce thrust from the direct impulse of exhaust gases are often called turbojets, whereas those that generate thrust with the addition of a ducted fan are often called turbofans or (rarely) fan-jets.

Gas turbines are also used in many liquid propellant rockets, the gas turbines are used to power a turbopump to permit the use of lightweight, low pressure tanks, which reduce the empty weight of the rocket.

Turboprop Engines

A turboprop engine is a turbine engine which drives an aircraft propeller using a reduction gear. Turboprop engines are used on small aircraft such as the general-aviation Cessna 208 Caravan and Embraer EMB 312 Tucano military trainer, medium-sized commuter aircraft such as the Bombardier Dash 8 and large aircraft such as the Airbus A400M transport and the 60 year-old Tupolev Tu-95 strategic bomber.

Aeroderivative Gas Turbines

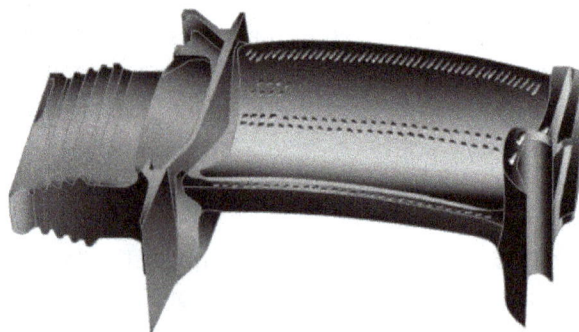

Diagram of a high-pressure film cooled turbine blade

Aeroderivatives are also used in electrical power generation due to their ability to be shut down, and handle load changes more quickly than industrial machines. They are also used in the marine industry to reduce weight. The General Electric LM2500, General Electric LM6000, Rolls-Royce RB211 and Rolls-Royce Avon are common models of this type of machine.

Amateur Gas Turbines

Increasing numbers of gas turbines are being used or even constructed by amateurs.

In its most straightforward form, these are commercial turbines acquired through military surplus or scrapyard sales, then operated for display as part of the hobby of engine collecting. In its most extreme form, amateurs have even rebuilt engines beyond professional repair and then used them to compete for the Land Speed Record.

The simplest form of self-constructed gas turbine employs an automotive turbocharger as the core component. A combustion chamber is fabricated and plumbed between the compressor and turbine sections.

More sophisticated turbojets are also built, where their thrust and light weight are sufficient to power large model aircraft. The Schreckling design constructs the entire engine from raw materials, including the fabrication of a centrifugal compressor wheel from plywood, epoxy and wrapped carbon fibre strands.

Several small companies now manufacture small turbines and parts for the amateur. Most turbojet-powered model aircraft are now using these commercial and semi-commercial microturbines, rather than a Schreckling-like home-build.

Auxiliary Power Units

APUs are small gas turbines designed to supply auxiliary power to larger, mobile, machines such as an aircraft. They supply:

- compressed air for air conditioning and ventilation,
- compressed air start-up power for larger jet engines,
- mechanical (shaft) power to a gearbox to drive shafted accessories or to start large jet engines, and
- electrical, hydraulic and other power-transmission sources to consuming devices remote from the APU.

Industrial Gas Turbines for Power Generation

Industrial gas turbines differ from aeronautical designs in that the frames, bearings, and blading are of heavier construction. They are also much more closely integrated with the devices they power— often an electric generator—and the secondary-energy equipment that is used to recover residual energy (largely heat).

They range in size from portable mobile plants to large, complex systems weighing more than a hundred tonnes housed in purpose-built buildings. When the gas turbine is used solely for shaft power, its thermal efficiency is about 30%. However, it may be cheaper to buy electricity than to generate it. Therefore, many engines are used in CHP (Combined Heat and Power) configurations that can be small enough to be integrated into portable container configurations.

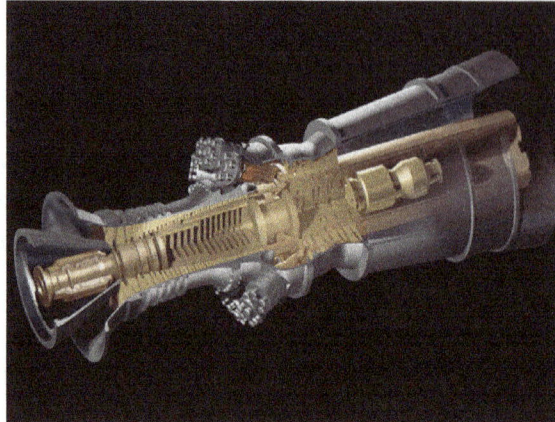

GE H series power generation gas turbine: in combined cycle configuration, this 480-megawatt unit has a rated thermal efficiency of 60%

Gas turbines can be particularly efficient when waste heat from the turbine is recovered by a heat recovery steam generator to power a conventional steam turbine in a combined cycle configuration. The 605 MW General Electric 9HA achieved a 62.22% efficiency rate with temperatures as high as 1,540 °C (2,800 °F). Aeroderivative gas turbines can also be used in combined cycles, leading to a higher efficiency, but it will not be as high as a specifically designed industrial gas turbine. They can also be run in a cogeneration configuration: the exhaust is used for space or water heating, or drives an absorption chiller for cooling the inlet air and increase the power output, technology known as Turbine Inlet Air Cooling.

Another significant advantage is their ability to be turned on and off within minutes, supplying power during peak, or unscheduled, demand. Since single cycle (gas turbine only) power plants are less efficient than combined cycle plants, they are usually used as peaking power plants, which operate anywhere from several hours per day to a few dozen hours per year—depending on the electricity demand and the generating capacity of the region. In areas with a shortage of base-load and load following power plant capacity or with low fuel costs, a gas turbine powerplant may regularly operate most hours of the day. A large single-cycle gas turbine typically produces 100 to 400 megawatts of electric power and has 35–40% thermal efficiency.

Industrial Gas Turbines for Mechanical Drive

Industrial gas turbines that are used solely for mechanical drive or used in collaboration with a recovery steam generator differ from power generating sets in that they are often smaller and feature a dual shaft design as opposed to single shaft. The power range varies from 1 megawatt up to 50 megawatts. These engines are connected directly or via a gearbox to either a pump or compressor assembly. The majority of installations are used within the oil and gas industries. Mechanical drive applications increase efficiency by around 2%.

Oil and Gas platforms require these engines to drive compressors to inject gas into the wells to force oil up via another bore, or to compress the gas for transportation. They're also often used to provide power for the platform. These platforms don't need to use the engine in collaboration with a CHP system due to getting the gas at an extremely reduced cost (often free from burn off gas). The same companies use pump sets to drive the fluids to land and across pipelines in various intervals.

Compressed Air Energy Storage

One modern development seeks to improve efficiency in another way, by separating the compressor and the turbine with a compressed air store. In a conventional turbine, up to half the generated power is used driving the compressor. In a compressed air energy storage configuration, power, perhaps from a wind farm or bought on the open market at a time of low demand and low price, is used to drive the compressor, and the compressed air released to operate the turbine when required.

Turboshaft Engines

Turboshaft engines are often used to drive compression trains (for example in gas pumping stations or natural gas liquefaction plants) and are used to power almost all modern helicopters. The primary shaft bears the compressor and the high speed turbine (often referred to as the *Gas Generator*), while a second shaft bears the low-speed turbine (a *power turbine* or *free-wheeling turbine* on helicopters, especially, because the gas generator turbine spins separately from the power turbine). In effect the separation of the gas generator, by a fluid coupling (the hot energy-rich combustion gases), from the power turbine is analogous to an automotive transmission's fluid coupling. This arrangement is used to increase power-output flexibility with associated highly-reliable control mechanisms.

Radial Gas Turbines

In 1963, Jan Mowill initiated the development at Kongsberg Våpenfabrikk in Norway. Various successors have made good progress in the refinement of this mechanism. Owing to a configuration that keeps heat away from certain bearings the durability of the machine is improved while the radial turbine is well matched in speed requirement.

Scale Jet Engines

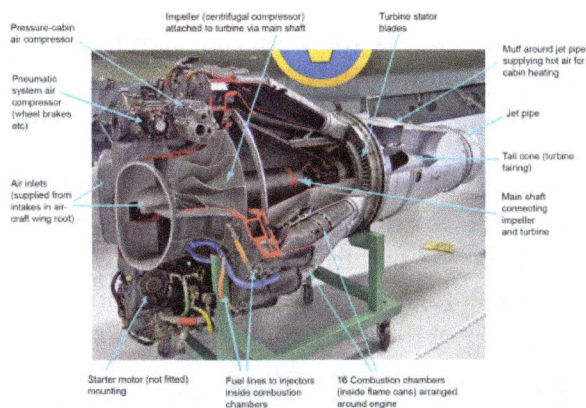

Scale jet engines are scaled down versions of this early full scale engine

Also known as miniature gas turbines or micro-jets.

With this in mind the pioneer of modern Micro-Jets, Kurt Schreckling, produced one of the world's first Micro-Turbines, the FD3/67. This engine can produce up to 22 newtons of thrust, and can be built by most mechanically minded people with basic engineering tools, such as a metal lathe.

Microturbines

Also known as:

- Turbo alternators
- Turbogenerator

Microturbines are becoming widespread in distributed power and combined heat and power applications, and are very promising for powering hybrid electric vehicles. They range from hand held units producing less than a kilowatt, to commercial sized systems that produce tens or hundreds of kilowatts. Basic principles of microturbine are based on micro-combustion.

Part of their claimed success is said to be due to advances in electronics, which allows unattended operation and interfacing with the commercial power grid. Electronic power switching technology eliminates the need for the generator to be synchronized with the power grid. This allows the generator to be integrated with the turbine shaft, and to double as the starter motor.

Microturbine systems have many claimed advantages over reciprocating engine generators, such as higher power-to-weight ratio, low emissions and few, or just one, moving part. Advantages are that microturbines may be designed with foil bearings and air-cooling operating without lubricating oil, coolants or other hazardous materials. Nevertheless, reciprocating engines overall are still cheaper when all factors are considered. Microturbines also have a further advantage of having the majority of the waste heat contained in the relatively high temperature exhaust making it simpler to capture, whereas the waste heat of reciprocating engines is split between its exhaust and cooling system.

However, reciprocating engine generators are quicker to respond to changes in output power requirement and are usually slightly more efficient, although the efficiency of microturbines is increasing. Microturbines also lose more efficiency at low power levels than reciprocating engines.

Reciprocating engines typically use simple motor oil (journal) bearings. Full-size gas turbines often use ball bearings. The 1000 °C temperatures and high speeds of microturbines make oil lubrication and ball bearings impractical; they require air bearings or possibly magnetic bearings.

When used in extended range electric vehicles the static efficiency drawback is irrelevant, since the gas turbine can be run at or near maximum power, driving an alternator to produce electricity either for the wheel motors, or for the batteries, as appropriate to speed and battery state. The batteries act as a "buffer" (energy storage) in delivering the required amount of power to the wheel motors, rendering throttle response of the gas turbine completely irrelevant.

There is, moreover, no need for a significant or variable-speed gearbox; turning an alternator at comparatively high speeds allows for a smaller and lighter alternator than would otherwise be the case. The superior power-to-weight ratio of the gas turbine and its fixed speed gearbox, allows for a much lighter prime mover than those in such hybrids as the Toyota Prius (which utilised a 1.8 litre petrol engine) or the Chevrolet Volt (which utilises a 1.4 litre petrol engine). This in turn allows a heavier weight of batteries to be carried, which allows for a longer electric-only range. Alternatively, the vehicle can use heavier types of batteries such as lead acid batteries (which are cheaper to buy) or safer types of batteries such as Lithium-Iron-Phosphate.

When gas turbines are used in extended-range electric vehicles, like those planned by Land-Rover/Range-Rover in conjunction with Bladon, or by Jaguar also in partnership with Bladon, the very poor throttling response (their high moment of rotational inertia) does not matter, because the gas turbine, which may be spinning at 100,000 rpm, is not directly, mechanically connected to the wheels. It was this poor throttling response that so bedevilled the 1950 Rover gas turbine-powered prototype motor car, which did not have the advantage of an intermediate electric drive train to provide sudden power spikes when demanded by the driver.

Gas turbines accept most commercial fuels, such as petrol, natural gas, propane, diesel, and kerosene as well as renewable fuels such as E85, biodiesel and biogas. However, when running on kerosene or diesel, starting sometimes requires the assistance of a more volatile product such as propane gas - although the new kero-start technology can allow even microturbines fuelled on kerosene to start without propane.

Microturbine designs usually consist of a single stage radial compressor, a single stage radial turbine and a recuperator. Recuperators are difficult to design and manufacture because they operate under high pressure and temperature differentials. Exhaust heat can be used for water heating, space heating, drying processes or absorption chillers, which create cold for air conditioning from heat energy instead of electric energy.

Typical microturbine efficiencies are 25 to 35%. When in a combined heat and power cogeneration system, efficiencies of greater than 80% are commonly achieved.

MIT started its millimeter size turbine engine project in the middle of the 1990s when Professor of Aeronautics and Astronautics Alan H. Epstein considered the possibility of creating a personal turbine which will be able to meet all the demands of a modern person's electrical needs, just as a large turbine can meet the electricity demands of a small city.

Problems have occurred with heat dissipation and high-speed bearings in these new microturbines. Moreover, their expected efficiency is a very low 5-6%. According to Professor Epstein, current commercial Li-ion rechargeable batteries deliver about 120-150 W·h/kg. MIT's millimeter size turbine will deliver 500-700 W·h/kg in the near term, rising to 1200-1500 W·h/kg in the longer term.

A similar microturbine built in Belgium has a rotor diameter of 20 mm and is expected to produce about 1000 W.

External Combustion

Most gas turbines are internal combustion engines but it is also possible to manufacture an external combustion gas turbine which is, effectively, a turbine version of a hot air engine. Those systems are usually indicated as EFGT (Externally Fired Gas Turbine) or IFGT (Indirectly Fired Gas Turbine).

External combustion has been used for the purpose of using pulverized coal or finely ground bio-mass (such as sawdust) as a fuel. In the indirect system, a heat exchanger is used and only clean air with no combustion products travels through the power turbine. The thermal efficiency is lower in the indirect type of external combustion; however, the turbine blades are not subjected to combustion products and much lower quality (and therefore cheaper) fuels are able to be used.

When external combustion is used, it is possible to use exhaust air from the turbine as the primary combustion air. This effectively reduces global heat losses, although heat losses associated with the combustion exhaust remain inevitable.

Closed-cycle gas turbines based on helium or supercritical carbon dioxide also hold promise for use with future high temperature solar and nuclear power generation.

In Surface Vehicles

The 1967 *STP Oil Treatment Special* on display at the Indianapolis Motor Speedway Hall of Fame Museum, with the Pratt & Whitney gas turbine shown

A 1968 Howmet TX, the only turbine-powered race car to have won a race

Gas turbines are often used on ships, locomotives, helicopters, tanks, and to a lesser extent, on cars, buses, and motorcycles.

A key advantage of jets and turboprops for aeroplane propulsion - their superior performance at high altitude compared to piston engines, particularly naturally aspirated ones - is irrelevant in most automobile applications. Their power-to-weight advantage, though less critical than for aircraft, is still important.

Gas turbines offer a high-powered engine in a very small and light package. However, they are not as responsive and efficient as small piston engines over the wide range of RPMs and powers needed in vehicle applications. In series hybrid vehicles, as the driving electric motors are mechanically detached from the electricity generating engine, the responsiveness, poor performance at low speed and low efficiency at low output problems are much less important. The turbine can be run at optimum speed for its power output, and batteries and ultracapacitors can supply power as needed, with the engine cycled on and off to run it only at high efficiency. The emergence of the continuously variable transmission may also alleviate the responsiveness problem.

Turbines have historically been more expensive to produce than piston engines, though this is partly because piston engines have been mass-produced in huge quantities for decades, while small gas turbine engines are rarities; however, turbines are mass-produced in the closely related form of the turbocharger.

The turbocharger is basically a compact and simple free shaft radial gas turbine which is driven by the piston engine's exhaust gas. The centripetal turbine wheel drives a centrifugal compressor wheel through a common rotating shaft. This wheel supercharges the engine air intake to a degree that can be controlled by means of a wastegate or by dynamically modifying the turbine housing's geometry (as in a VGT turbocharger). It mainly serves as a power recovery device which converts a great deal of otherwise wasted thermal and kinetic energy into engine boost.

Turbo-compound engines (actually employed on some trucks) are fitted with blow down turbines which are similar in design and appearance to a turbocharger except for the turbine shaft being mechanically or hydraulically connected to the engine's crankshaft instead of to a centrifugal compressor, thus providing additional power instead of boost. While the turbocharger is a pressure turbine, a power recovery turbine is a velocity one.

Passenger Road Vehicles (Cars, Bikes, and Buses)

A number of experiments have been conducted with gas turbine powered automobiles, the largest by Chrysler. More recently, there has been some interest in the use of turbine engines for hybrid electric cars. For instance, a consortium led by micro gas turbine company Bladon Jets has secured investment from the Technology Strategy Board to develop an Ultra Lightweight Range Extender (ULRE) for next generation electric vehicles. The objective of the consortium, which includes luxury car maker Jaguar Land Rover and leading electrical machine company SR Drives, is to produce the world's first commercially viable - and environmentally friendly - gas turbine generator designed specifically for automotive applications.

The common turbocharger for gasoline or diesel engines is also a turbine derivative.

Concept Cars

The first serious investigation of using a gas turbine in cars was in 1946 when two engineers, Robert Kafka and Robert Engerstein of Carney Associates, a New York engineering firm, came up with the concept where a unique compact turbine engine design would provide power for a rear wheel drive car. After an article appeared in *Popular Science*, there was no further work, beyond the paper stage.

The 1950 Rover JET1

In 1950, designer F.R. Bell and Chief Engineer Maurice Wilks from British car manufacturers Rover unveiled the first car powered with a gas turbine engine. The two-seater JET1 had the engine positioned behind the seats, air intake grilles on either side of the car, and exhaust outlets on the top of the tail. During tests, the car reached top speeds of 140 km/h (87 mph), at a turbine speed of 50,000 rpm. The car ran on petrol, paraffin (kerosene) or diesel oil, but fuel consumption problems proved insurmountable for a production car. It is on display at the London Science Museum.

A French turbine powered car, the Socema-Gregoire, was displayed at the October 1952 Paris Auto Show. It was designed by the French engineer Jean-Albert Grégoire.

Firebird I

The first turbine powered car built in the US was the GM Firebird I which began evaluations in 1953. While photos of the Firebird I may suggest that the jet turbine's thrust propelled the car like an aircraft, the turbine in fact drove the rear wheels. The Firebird 1 was never meant as a serious commercial passenger car and was solely built for testing & evaluation as well as public relation purposes.

Starting in 1954 with a modified Plymouth, the American car manufacturer Chrysler demonstrated several prototype gas turbine-powered cars from the early 1950s through the early 1980s. Chrysler built fifty Chrysler Turbine Cars in 1963 and conducted the only consumer trial of gas turbine-powered cars. Each of their turbines employed a unique rotating recuperator, referred to as a regenerator that increased efficiency.

In 1954 FIAT unveiled a concept car with a turbine engine, called Fiat Turbina. This vehicle, looking like an aircraft with wheels, used a unique combination of both jet thrust and the engine driving the wheels. Speeds of 282 km/h (175 mph) were claimed.

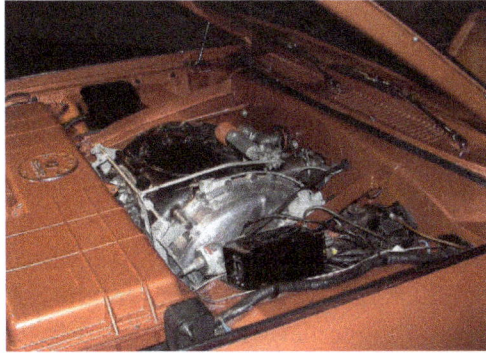

Engine compartment of a Chrysler 1963 Turbine car

The original General Motors Firebird was a series of concept cars developed for the 1953, 1956 and 1959 Motorama auto shows, powered by gas turbines.

As a result of the U.S. Clean Air Act Amendments of 1970, research was funded to developing automotive gas turbine technology. Design concepts and vehicles were conducted by Chrysler, General Motors, Ford (in collaboration with AiResearch), and American Motors (in conjunction with Williams Research). Long-term tests were conducted evaluate comparable cost efficiency. Several AMC Hornets were powered by a small Williams regenerative gas turbines weighing 250 lb (113 kg) and producing 80 hp (60 kW; 81 PS) at 4450 rpm.

Toyota demonstrated several gas turbine powered concept cars, such as the Century gas turbine hybrid in 1975, the Sports 800 Gas Turbine Hybrid in 1979 and the GTV in 1985. No production vehicles were made. The GT24 engine was exhibited in 1977 without a vehicle.

In the early 1990s Volvo introduced the Volvo Environmental Concept Car(ECC) which was a gas turbine powered hybrid car.

In 1993 General Motors introduced the first commercial gas turbine powered hybrid vehicle—as a limited production run of the EV-1 series hybrid. A Williams International 40 kW turbine drove an alternator which powered the battery-electric powertrain. The turbine design included a recuperator. Later on in 2006 GM went into the EcoJet concept car project with Jay Leno.

At the 2010 Paris Motor Show Jaguar demonstrated its Jaguar C-X75 concept car. This electrically powered supercar has a top speed of 204 mph (328 km/h) and can go from 0 to 62 mph (0 to 100 km/h) in 3.4 seconds. It uses Lithium-ion batteries to power 4 electric motors which combine to produce some 780 bhp. It will do 68 miles (109 km) on a single charge of the batteries, but in addition it uses a pair of Bladon Micro Gas Turbines to re-charge the batteries extending the range to 560 miles (900 km).

Racing Cars

The first race car (in concept only) fitted with a turbine was in 1955 by a US Air Force group as a hobby project with a turbine loaned them by Boeing and a race car owned by Firestone Tire & Rubber company. The first race car fitted with a turbine for the goal of actual racing was by Rover and the BRM Formula One team joined forces to produce the Rover-BRM, a gas turbine powered coupe, which entered the 1963 24 Hours of Le Mans, driven by Graham Hill and Richie Ginther. It averaged 107.8 mph (173.5 km/h) and had a top speed of 142 mph (229 km/h). American Ray Heppenstall joined Howmet Corporation and McKee Engineering together to develop their own gas turbine sports car in 1968, the Howmet TX, which ran several American and European events, including two wins, and also participated in the 1968 24 Hours of Le Mans. The cars used Continental gas turbines, which eventually set six FIA land speed records for turbine-powered cars.

For open wheel racing, 1967's revolutionary STP-Paxton Turbocar fielded by racing and entrepreneurial legend Andy Granatelli and driven by Parnelli Jones nearly won the Indianapolis 500; the Pratt & Whitney ST6B-62 powered turbine car was almost a lap ahead of the second place car when a gearbox bearing failed just three laps from the finish line. The next year the STP Lotus 56 turbine car won the Indianapolis 500 pole position even though new rules restricted the air intake dramatically. In 1971 Lotus principal Colin Chapman introduced the Lotus 56B F1 car, powered by a Pratt & Whitney STN 6/76 gas turbine. Chapman had a reputation of building radical championship-winning cars, but had to abandon the project because there were too many problems with turbo lag.

Buses

The arrival of the Capstone Microturbine has led to several hybrid bus designs, starting with HEV-1 by AVS of Chattanooga, Tennessee in 1999, and closely followed by Ebus and ISE Research in California, and DesignLine Corporation in New Zealand (and later the United States). AVS turbine hybrids were plagued with reliability and quality control problems, resulting in liquidation of AVS in 2003. The most successful design by Designline is now operated in 5 cities in 6 countries, with over 30 buses in operation worldwide, and order for several hundred being delivered to Baltimore, and NYC.

Brescia Italy is using serial hybrid buses powered by microturbines on routes through the historical sections of the city.

Motorcycles

The MTT Turbine Superbike appeared in 2000 (hence the designation of Y2K Superbike by MTT) and is the first production motorcycle powered by a turbine engine - specifically, a Rolls-Royce Allison model 250 turboshaft engine, producing about 283 kW (380 bhp). Speed-tested to 365 km/h or 227 mph (according to some stories, the testing team ran out of road during the test), it holds the Guinness World Record for most powerful production motorcycle and most expensive production motorcycle, with a price tag of US$185,000.

Trains

Several locomotive classes have been powered by gas turbines, the most recent incarnation being Bombardier's JetTrain.

Tanks

The German Army's development division, the Heereswaffenamt (Army Ordnance Board), studied a number of gas turbine engines for use in tanks starting in mid-1944. The first gas turbine engines used for armoured fighting vehicle GT 101 was installed in the Panther tank. The second use of a gas turbine in an armoured fighting vehicle was in 1954 when a unit, PU2979, specifically developed for tanks by C. A. Parsons & Co., was installed and trialled in a British Conqueror tank.

The Stridsvagn 103 was developed in the 1950s and was the first mass-produced main battle tank to use a turbine engine. Since then, gas turbine engines have been used as APUs in some tanks and as main powerplants in Soviet/Russian T-80s and U.S. M1 Abrams tanks, among others. They are lighter and smaller than diesels at the same sustained power output but the models installed to date are less fuel efficient than the equivalent diesel, especially at idle, requiring more fuel to achieve the same combat range. Successive models of M1 have addressed this problem with battery packs or secondary generators to power the tank's systems while stationary, saving fuel by reducing the need to idle the main turbine. T-80s can mount three large external fuel drums to extend their range. Russia has stopped production of the T-80 in favour of the diesel-powered T-90 (based on the T-72), while Ukraine has developed the diesel-powered T-80UD and T-84 with nearly the power of the gas-turbine tank. The French Leclerc MBT's diesel powerplant features the "Hyperbar" hybrid supercharging system, where the engine's turbocharger is completely replaced with a small gas turbine which also works as an assisted diesel exhaust turbocharger, enabling engine RPM-independent boost level control and a higher peak boost pressure to be reached (than with ordinary turbochargers). This system allows a smaller displacement and lighter engine to be used as the tank's powerplant and effectively removes turbo lag. This special gas turbine/turbocharger can also work independently from the main engine as an ordinary APU.

Marines from 1st Tank Battalion load a Honeywell AGT1500 multi-fuel turbine back into an M1 Abrams tank at Camp Coyote, Kuwait, February 2003

A turbine is theoretically more reliable and easier to maintain than a piston engine, since it has a simpler construction with fewer moving parts but in practice turbine parts experience a higher wear rate due to their higher working speeds. The turbine blades are highly sensitive to dust and fine sand, so that in desert operations air filters have to be fitted and changed several times daily. An improperly fitted filter, or a bullet or shell fragment that punctures the filter, can damage the engine. Piston engines (especially if turbocharged) also need well-maintained filters, but they are more resilient if the filter does fail.

Like most modern diesel engines used in tanks, gas turbines are usually multi-fuel engines.

Marine Applications

Naval

Gas turbines are used in many naval vessels, where they are valued for their high power-to-weight ratio and their ships' resulting acceleration and ability to get underway quickly.

The Gas turbine from MGB 2009

The first gas-turbine-powered naval vessel was the Royal Navy's Motor Gun Boat *MGB 2009* (formerly *MGB 509*) converted in 1947. Metropolitan-Vickers fitted their F2/3 jet engine with a power turbine. The Steam Gun Boat *Grey Goose* was converted to Rolls-Royce gas turbines in 1952 and operated as such from 1953. The Bold class Fast Patrol Boats *Bold Pioneer* and *Bold Pathfinder* built in 1953 were the first ships created specifically for gas turbine propulsion.

The first large scale, partially gas-turbine powered ships were the Royal Navy's Type 81 (Tribal class) frigates with combined steam and gas powerplants. The first, HMS *Ashanti* was commissioned in 1961.

The German Navy launched the first *Köln*-class frigate in 1961 with 2 Brown, Boveri & Cie gas turbines in the world's first combined diesel and gas propulsion system.

The Danish Navy had 6 *Søløven*-class torpedo boats (the export version of the British Brave class fast patrol boat) in service from 1965 to 1990, which had 3 Bristol Proteus (later RR Proteus) Marine Gas Turbines rated at 9,510 kW (12,750 shp) combined, plus two General Motors Diesel engines, rated at 340 kW (460 shp), for better fuel economy at slower speeds. And they also produced 10 Willemoes Class Torpedo / Guided Missile boats (in service from 1974 to 2000) which had 3 Rolls Royce Marine Proteus Gas Turbines also rated at 9,510 kW (12,750 shp), same as the Søløven-class boats, and 2 General Motors Diesel Engines, rated at 600 kW (800 shp), also for improved fuel economy at slow speeds.

The Swedish Navy produced 6 Spica-class torpedo boats between 1966 and 1967 powered by 3 Bristol Siddeley Proteus 1282 turbines, each delivering 3,210 kW (4,300 shp). They were later joined by 12 upgraded Norrköping class ships, still with the same engines. With their aft torpedo tubes replaced by antishipping missiles they served as missile boats until the last was retired in 2005.

The Finnish Navy commissioned two *Turunmaa*-class corvettes, *Turunmaa* and *Karjala*, in 1968. They were equipped with one 16,410 kW (22,000 shp) Rolls-Royce Olympus TM1 gas turbine and three Wärtsilä marine diesels for slower speeds. They were the fastest vessels in the Finnish Navy; they regularly achieved speeds of 35 knots, and 37.3 knots during sea trials. The *Turunmaas* were paid off in 2002. *Karjala* is today a museum ship in Turku, and *Turunmaa* serves as a floating machine shop and training ship for Satakunta Polytechnical College.

The next series of major naval vessels were the four Canadian *Iroquois*-class helicopter carrying destroyers first commissioned in 1972. They used 2 ft-4 main propulsion engines, 2 ft-12 cruise engines and 3 Solar Saturn 750 kW generators.

An LM2500 gas turbine on USS *Ford*

The first U.S. gas-turbine powered ship was the U.S. Coast Guard's *Point Thatcher*, a cutter commissioned in 1961 that was powered by two 750 kW (1,000 shp) turbines utilizing controllable-pitch propellers. The larger *Hamilton*-class High Endurance Cutters, was the first class of larger cutters to utilize gas turbines, the first of which (USCGC *Hamilton*) was commissioned in 1967. Since then, they have powered the U.S. Navy's *Oliver Hazard Perry*-class frigates, *Spruance* and *Arleigh Burke*-class destroyers, and *Ticonderoga*-class guided missile cruisers. USS *Makin Island*, a modified *Wasp*-class amphibious assault ship, is to be the Navy's first amphibious assault ship powered by gas turbines. The marine gas turbine operates in a more corrosive atmosphere due to presence of sea salt in air and fuel and use of cheaper fuels.

Civilian Maritime

Up to the late 1940s much of the progress on marine gas turbines all over the world took place in design offices and engine builder's workshops and development work was led by the British Royal Navy and other Navies. While interest in the gas turbine for marine purposes, both naval and mercantile, continued to increase, the lack of availability of the results of operating experience on early gas turbine projects limited the number of new ventures on seagoing commercial vessels being embarked upon. In 1951, the Diesel-electric oil tanker *Auris*, 12,290 Deadweight tonnage (DWT) was used to obtain operating experience with a main propulsion gas turbine under service conditions at sea and so became the first ocean-going merchant ship to be powered by a gas turbine. Built by Hawthorn Leslie at Hebburn-on-Tyne, UK, in accordance with plans and specifications drawn up by the Anglo-Saxon Petroleum Company and launched on the UK's Princess Elizabeth's 21st birthday in 1947, the ship was designed with an engine room layout that would allow for the experimental use of heavy fuel in one of its high-speed engines, as well as the future substitution of one of its diesel engines by a gas turbine. The *Auris* operated commercially as a tanker for three-and-a-half years with a diesel-electric propulsion unit as originally commissioned, but in 1951 one of its four 824 kW (1,105 bhp) diesel engines – which were known as "Faith", "Hope", "Charity" and "Prudence" - was replaced by the world's first marine gas turbine engine, a 890 kW (1,200 bhp) open-cycle gas turbo-alternator built by British Thomson-Houston Company in Rugby. Following successful sea trials off the Northumbrian coast, the *Auris* set sail from Hebburn-on-Tyne in October

1951 bound for Port Arthur in the US and then Curacao in the southern Caribbean returning to Avonmouth after 44 days at sea, successfully completing her historic trans-Atlantic crossing. During this time at sea the gas turbine burnt diesel fuel and operated without an involuntary stop or mechanical difficulty of any kind. She subsequently visited Swansea, Hull, Rotterdam, Oslo and Southampton covering a total of 13,211 nautical miles. The *Auris* then had all of its power plants replaced with a 3,910 kW (5,250 shp) directly coupled gas turbine to become the first civilian ship to operate solely on gas turbine power.

Despite the success of this early experimental voyage the gas turbine was not to replace the diesel engine as the propulsion plant for large merchant ships. At constant cruising speeds the diesel engine simply had no peer in the vital area of fuel economy. The gas turbine did have more success in Royal Navy ships and the other naval fleets of the world where sudden and rapid changes of speed are required by warships in action.

The United States Maritime Commission were looking for options to update WWII Liberty ships, and heavy-duty gas turbines were one of those selected. In 1956 the *John Sergeant* was lengthened and equipped with a General Electric 4,900 kW (6,600 shp) HD gas turbine with exhaust-gas regeneration, reduction gearing and a variable-pitch propeller. It operated for 9,700 hours using residual fuel(Bunker C) for 7,000 hours. Fuel efficiency was on a par with steam propulsion at 0.318 kg/kW (0.523 lb/hp) per hour, and power output was higher than expected at 5,603 kW (7,514 shp) due to the ambient temperature of the North Sea route being lower than the design temperature of the gas turbine. This gave the ship a speed capability of 18 knots, up from 11 knots with the original power plant, and well in excess of the 15 knot targeted. The ship made its first transatlantic crossing with an average speed of 16.8 knots, in spite of some rough weather along the way. Suitable Bunker C fuel was only available at limited ports because the quality of the fuel was of a critical nature. The fuel oil also had to be treated on board to reduce contaminants and this was a labor-intensive process that was not suitable for automation at the time. Ultimately, the variable-pitch propeller, which was of a new and untested design, ended the trial, as three consecutive annual inspections revealed stress-cracking. This did not reflect poorly on the marine-propulsion gas-turbine concept though, and the trial was a success overall. The success of this trial opened the way for more development by GE on the use of HD gas turbines for marine use with heavy fuels. The *John Sergeant* was scrapped in 1972 at Portsmouth PA.

Between 1971 and 1981, Seatrain Lines operated a scheduled container service between ports on the eastern seaboard of the United States and ports in northwest Europe across the North Atlantic with four container ships of 26,000 tonnes DWT. Those ships were powered by twin Pratt & Whitney gas turbines of the FT 4 series. The four ships in the class were named *Euroliner, Eurofreighter, Asialiner* and *Asiafreighter*. Following the dramatic Organization of the Petroleum Exporting Countries (OPEC) price increases of the mid-1970s, operations were constrained by rising fuel costs. Some modification of the engine systems on those ships was undertaken to permit the burning of a lower grade of fuel (i.e., marine diesel). Reduction of fuel costs was successful using a different untested fuel in a marine gas turbine but maintenance costs increased with the fuel change. After 1981 the ships were sold and refitted with, what at the time, was more economical diesel-fueled engines but the increased engine size reduced cargo space.

Boeing Jetfoil 929-100-007 *Urzela* of TurboJET

Boeing launched its first passenger-carrying waterjet-propelled hydrofoil Boeing 929, in April 1974. Those ships were powered by twin Allison gas turbines of the KF-501 series.

The first passenger ferry to use a gas turbine was the GTS *Finnjet*, built in 1977 and powered by two Pratt & Whitney FT 4C-1 DLF turbines, generating 55,000 kW (74,000 shp) and propelling the ship to a speed of 31 knots. However, the Finnjet also illustrated the shortcomings of gas turbine propulsion in commercial craft, as high fuel prices made operating her unprofitable. After four years of service additional diesel engines were installed on the ship to reduce running costs during the off-season. The Finnjet was also the first ship with a Combined diesel-electric and gas propulsion. Another example of commercial usage of gas turbines in a passenger ship is Stena Line's HSS class fastcraft ferries. HSS 1500-class *Stena Explorer*, *Stena Voyager* and *Stena Discovery* vessels use combined gas and gas setups of twin GE LM2500 plus GE LM1600 power for a total of 68,000 kW (91,000 shp). The slightly smaller HSS 900-class *Stena Carisma*, uses twin ABB–STAL (sv) GT35 turbines rated at 34,000 kW (46,000 shp) gross. The *Stena Discovery* was withdrawn from service in 2007, another victim of too high fuel costs.

In July 2000 the *Millennium* became the first cruise ship to be propelled by gas turbines, in a Combined Gas and Steam Turbine configuration. The liner RMS Queen Mary 2 uses a Combined Diesel and Gas Turbine configuration.

In marine racing applications the 2010 C5000 Mystic catamaran Miss GEICO uses two Lycoming T-55 turbines for its power system.

Advances in Technology

Gas turbine technology has steadily advanced since its inception and continues to evolve. Development is actively producing both smaller gas turbines and more powerful and efficient engines. Aiding in these advances are computer based design (specifically CFD and finite element analysis) and the development of advanced materials: Base materials with superior high temperature strength (e.g., single-crystal superalloys that exhibit yield strength anomaly) or thermal barrier coatings that protect the structural material from ever higher temperatures. These advances allow higher compression ratios and turbine inlet temperatures, more efficient combustion and better cooling of engine parts.

Computational Fluid Dynamics (CFD) has contributed to substantial improvements in the performance and efficiency of Gas Turbine engine components through enhanced understanding of the complex viscous flow and heat transfer phenomena involved. For this reason, CFD is one of the key computational tool used in Design & development of gas turbine engines.

The simple-cycle efficiencies of early gas turbines were practically doubled by incorporating inter-cooling, regeneration (or recuperation), and reheating. These improvements, of course, come at the expense of increased initial and operation costs, and they cannot be justified unless the decrease in fuel costs offsets the increase in other costs. The relatively low fuel prices, the general desire in the industry to minimize installation costs, and the tremendous increase in the simple-cycle efficiency to about 40 percent left little desire for opting for these modifications.

On the emissions side, the challenge is to increase turbine inlet temperatures while at the same time reducing peak flame temperature in order to achieve lower NOx emissions and meet the latest emission regulations. In May 2011, Mitsubishi Heavy Industries achieved a turbine inlet temperature of 1,600 °C on a 320 megawatt gas turbine, and 460 MW in gas turbine combined-cycle power generation applications in which gross thermal efficiency exceeds 60%.

Compliant foil bearings were commercially introduced to gas turbines in the 1990s. These can withstand over a hundred thousand start/stop cycles and have eliminated the need for an oil system. The application of microelectronics and power switching technology have enabled the development of commercially viable electricity generation by micro turbines for distribution and vehicle propulsion.

Advantages and Disadvantages

The following are advantages and disadvantages of gas-turbine engines:

Advantages

- Very high power-to-weight ratio, compared to reciprocating engines
- Smaller than most reciprocating engines of the same power rating
- Moves in one direction only, with far less vibration than a reciprocating engine
- Fewer moving parts than reciprocating engines implies lower maintenance cost.
- Greater reliability, particularly in applications where sustained high power output is required
- Waste heat is dissipated almost entirely in the exhaust. This results in a high temperature exhaust stream that is very usable for boiling water in a combined cycle, or for cogeneration
- Low operating pressures
- High operation speeds
- Low lubricating oil cost and consumption
- Can run on a wide variety of fuels

- Very low toxic emissions of CO and HC due to excess air, complete combustion and no "quench" of the flame on cold surfaces

Disadvantages

- Cost is very high
- Less efficient than reciprocating engines at idle speed
- Longer startup than reciprocating engines
- Less responsive to changes in power demand compared with reciprocating engines
- Characteristic whine can be hard to suppress

Testing

British, German, other national and international test codes are used to standardize the procedures and definitions used to test gas turbines. Selection of the test code to be used is an agreement between the purchaser and the manufacturer, and has some significance to the design of the turbine and associated systems. In the United States, ASME has produced several performance test codes on gas turbines. This includes ASME PTC 22-2014. These ASME performance test codes have gained international recognition and acceptance for testing gas turbines. The single most important and differentiating characteristic of ASME performance test codes, including PTC 22, is that the test uncertainty of the measurement indicates the quality of the test and is not to be used as a commercial tolerance.

Turbine

A turbine (from the Latin *turbo*, a vortex, related to the Greek, *tyrbē*, meaning "turbulence") is a rotary mechanical device that extracts energy from a fluid flow and converts it into useful work. A turbine is a turbomachine with at least one moving part called a rotor assembly, which is a shaft or drum with blades attached. Moving fluid acts on the blades so that they move and impart rotational energy to the rotor. Early turbine examples are windmills and waterwheels.

Gas, steam, and water turbines have a casing around the blades that contains and controls the working fluid. Credit for invention of the steam turbine is given both to British engineer Sir Charles Parsons (1854–1931) for invention of the reaction turbine, and to Swedish engineer Gustaf de Laval (1845–1913) for invention of the impulse turbine. Modern steam turbines frequently employ both reaction and impulse in the same unit, typically varying the degree of reaction and impulse from the blade root to its periphery.

The word "turbine" was coined in 1822 by the French mining engineer Claude Burdin from the Latin *turbo*, or vortex, in a memo, "Des turbines hydrauliques ou machines rotatoires à grande vitesse", which he submitted to the Académie royale des sciences in Paris. Benoit Fourneyron, a former student of Claude Burdin, built the first practical water turbine.

A steam turbine with the case opened.

Operation Theory

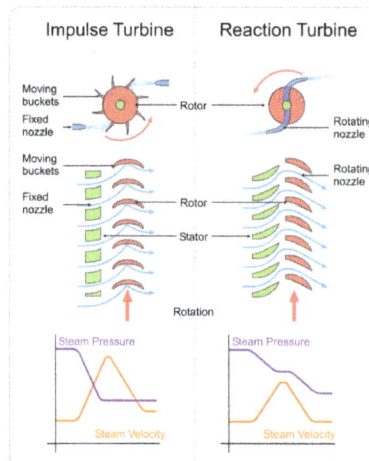

Schematic of impulse and reaction turbines, where the rotor is the rotating part, and the stator is the stationary part of the machine.

A working fluid contains potential energy (pressure head) and kinetic energy (velocity head). The fluid may be compressible or incompressible. Several physical principles are employed by turbines to collect this energy:

Impulse turbines change the direction of flow of a high velocity fluid or gas jet. The resulting impulse spins the turbine and leaves the fluid flow with diminished kinetic energy. There is no pressure change of the fluid or gas in the turbine blades (the moving blades), as in the case of a steam or gas turbine, all the pressure drop takes place in the stationary blades (the nozzles). Before reaching the turbine, the fluid's *pressure head* is changed to *velocity head* by accelerating the fluid with a nozzle. Pelton wheels and de Laval turbines use this process exclusively. Impulse turbines do not require a pressure casement around the rotor since the fluid jet is created by the nozzle prior to reaching the blades on the rotor. Newton's second law describes the transfer of energy for impulse turbines.

Reaction turbines develop torque by reacting to the gas or fluid's pressure or mass. The pressure of the gas or fluid changes as it passes through the turbine rotor blades. A pressure casement is needed to contain the working fluid as it acts on the turbine stage(s) or the turbine must be fully immersed in the fluid flow (such as with wind turbines). The casing contains and directs the working fluid and, for water turbines, maintains the suction imparted by the draft tube. Francis turbines and most steam turbines use this concept. For compressible working fluids, multiple turbine stages are usually used to harness the expanding gas efficiently. Newton's third law describes the transfer of energy for reaction turbines.

In the case of steam turbines, such as would be used for marine applications or for land-based electricity generation, a Parsons type reaction turbine would require approximately double the number of blade rows as a de Laval type impulse turbine, for the same degree of thermal energy conversion. Whilst this makes the Parsons turbine much longer and heavier, the overall efficiency of a reaction turbine is slightly higher than the equivalent impulse turbine for the same thermal energy conversion.

In practice, modern turbine designs use both reaction and impulse concepts to varying degrees whenever possible. Wind turbines use an airfoil to generate a reaction lift from the moving fluid and impart it to the rotor. Wind turbines also gain some energy from the impulse of the wind, by deflecting it at an angle. Turbines with multiple stages may utilize either reaction or impulse blading at high pressure. Steam turbines were traditionally more impulse but continue to move towards reaction designs similar to those used in gas turbines. At low pressure the operating fluid medium expands in volume for small reductions in pressure. Under these conditions, blading becomes strictly a reaction type design with the base of the blade solely impulse. The reason is due to the effect of the rotation speed for each blade. As the volume increases, the blade height increases, and the base of the blade spins at a slower speed relative to the tip. This change in speed forces a designer to change from impulse at the base, to a high reaction style tip.

Classical turbine design methods were developed in the mid 19th century. Vector analysis related the fluid flow with turbine shape and rotation. Graphical calculation methods were used at first. Formulae for the basic dimensions of turbine parts are well documented and a highly efficient machine can be reliably designed for any fluid flow condition. Some of the calculations are empirical or 'rule of thumb' formulae, and others are based on classical mechanics. As with most engineering calculations, simplifying assumptions were made.

Velocity triangles can be used to calculate the basic performance of a turbine stage. Gas exits the stationary turbine nozzle guide vanes at absolute velocity V_{a1}. The rotor rotates at velocity U. Relative to the rotor, the velocity of the gas as it impinges on the rotor entrance is V_{r1}. The gas is turned by the rotor and exits, relative to the rotor, at velocity V_{r2}. However, in absolute terms the rotor exit velocity is V_{a2}. The velocity triangles are constructed using these various velocity vectors. Velocity triangles can be constructed at any section through the blading (for example: hub, tip, midsection and so on) but are usually shown at the mean stage radius. Mean performance for the stage can be calculated from the velocity triangles, at this radius, using the Euler equation:

$$\Delta h = u \cdot \Delta v_w$$

Turbine inlet guide vanes of a turbojet

Hence:

$$\frac{\Delta h}{T} = \frac{u \cdot \Delta v_w}{T}$$

where:

Δh is the specific enthalpy drop across stage

T is the turbine entry total (or stagnation) temperature

u is the turbine rotor peripheral velocity

Δv_w is the change in whirl velocity

The turbine pressure ratio is a function of $\dfrac{\Delta h}{T}$ and the turbine efficiency.

Modern turbine design carries the calculations further. Computational fluid dynamics dispenses with many of the simplifying assumptions used to derive classical formulas and computer software facilitates optimization. These tools have led to steady improvements in turbine design over the last forty years.

The primary numerical classification of a turbine is its *specific speed*. This number describes the speed of the turbine at its maximum efficiency with respect to the power and flow rate. The specific speed is derived to be independent of turbine size. Given the fluid flow conditions and the desired shaft output speed, the specific speed can be calculated and an appropriate turbine design selected.

The specific speed, along with some fundamental formulas can be used to reliably scale an existing design of known performance to a new size with corresponding performance.

Off-design performance is normally displayed as a turbine map or characteristic.

Types

- Steam turbines are used for the generation of electricity in thermal power plants, such as plants using coal, fuel oil or nuclear fuel. They were once used to directly drive mechanical devices such as ships' propellers (for example the *Turbinia*, the first turbine-powered steam launch,) but most such applications now use reduction gears or an intermediate electrical step, where the turbine is used to generate electricity, which then powers an electric motor connected to the mechanical load. Turbo electric ship machinery was particularly popular in the period immediately before and during World War II, primarily due to a lack of sufficient gear-cutting facilities in US and UK shipyards.

- Gas turbines are sometimes referred to as turbine engines. Such engines usually feature an inlet, fan, compressor, combustor and nozzle (possibly other assemblies) in addition to one or more turbines.

- Transonic turbine. The gas flow in most turbines employed in gas turbine engines remains subsonic throughout the expansion process. In a transonic turbine the gas flow becomes supersonic as it exits the nozzle guide vanes, although the downstream velocities normally become subsonic. Transonic turbines operate at a higher pressure ratio than normal but are usually less efficient and uncommon.

- Contra-rotating turbines. With axial turbines, some efficiency advantage can be obtained if a downstream turbine rotates in the opposite direction to an upstream unit. However, the complication can be counter-productive. A contra-rotating steam turbine, usually known as the Ljungström turbine, was originally invented by Swedish Engineer Fredrik Ljungström (1875–1964) in Stockholm, and in partnership with his brother Birger Ljungström he obtained a patent in 1894. The design is essentially a multi-stage radial turbine (or pair of 'nested' turbine rotors) offering great efficiency, four times as large heat drop per stage as in the reaction (Parsons) turbine, extremely compact design and the type met particular success in back pressure power plants. However, contrary to other designs, large steam volumes are handled with difficulty and only a combination with axial flow turbines (DUREX) admits the turbine to be built for power greater than ca 50 MW. In marine applications only about 50 turbo-electric units were ordered (of which a considerable amount were finally sold to land plants) during 1917-19, and during 1920-22 a few turbo-mechanic not very successful units were sold. Only a few turbo-electric marine plants were still in use in the late 1960s (ss Ragne, ss Regin) while most land plants remain in use 2010.

- Statorless turbine. Multi-stage turbines have a set of static (meaning stationary) inlet guide vanes that direct the gas flow onto the rotating rotor blades. In a stator-less turbine the gas flow exiting an upstream rotor impinges onto a downstream rotor without an intermediate set of stator vanes (that rearrange the pressure/velocity energy levels of the flow) being encountered.

- Ceramic turbine. Conventional high-pressure turbine blades (and vanes) are made from nickel based alloys and often utilise intricate internal air-cooling passages to prevent the metal from overheating. In recent years, experimental ceramic blades have been manufactured and tested in gas turbines, with a view to increasing rotor inlet temperatures and/ or, possibly, eliminating air cooling. Ceramic blades are more brittle than their metallic counterparts, and carry a greater risk of catastrophic blade failure. This has tended to limit their use in jet engines and gas turbines to the stator (stationary) blades.

- Shrouded turbine. Many turbine rotor blades have shrouding at the top, which interlocks with that of adjacent blades, to increase damping and thereby reduce blade flutter. In large land-based electricity generation steam turbines, the shrouding is often complemented, especially in the long blades of a low-pressure turbine, with lacing wires. These wires pass through holes drilled in the blades at suitable distances from the blade root and are usually brazed to the blades at the point where they pass through. Lacing wires reduce blade flutter in the central part of the blades. The introduction of lacing wires substantially reduces the instances of blade failure in large or low-pressure turbines.

- Shroudless turbine. Modern practice is, wherever possible, to eliminate the rotor shrouding, thus reducing the centrifugal load on the blade and the cooling requirements.

- Bladeless turbine uses the boundary layer effect and not a fluid impinging upon the blades as in a conventional turbine.

- Water turbines

    - Pelton turbine, a type of impulse water turbine.

    - Francis turbine, a type of widely used water turbine.

    - Kaplan turbine, a variation of the Francis Turbine.

    - Turgo turbine, a modified form of the Pelton wheel.

    - Cross-flow turbine, also known as Banki-Michell turbine, or Ossberger turbine.

- Wind turbine. These normally operate as a single stage without nozzle and interstage guide vanes. An exception is the Éolienne Bollée, which has a stator and a rotor.

- Velocity compound "Curtis". Curtis combined the de Laval and Parsons turbine by using a set of fixed nozzles on the first stage or stator and then a rank of fixed and rotating blade rows, as in the Parsons or de Laval, typically up to ten compared with up to a hundred stages of a Parsons design. The overall efficiency of a Curtis design is less than that of either the Parsons or de Laval designs, but it can be satisfactorily operated through a much wider range of speeds, including successful operation at low speeds and at lower pressures, which made it ideal for use in ships' powerplant. In a Curtis arrangement, the entire heat drop in the steam takes place in the initial nozzle row and both the subsequent moving blade rows and stationary blade rows merely change the direction of the steam. Use of a small section of a Curtis arrangement, typically one nozzle section and two or three rows of moving blades, is usually termed a Curtis 'Wheel' and in this form, the Curtis found widespread use at sea as a 'governing stage' on many reaction and impulse turbines and turbine sets. This practice is still commonplace today in marine steam plant.

- Pressure compound multi-stage impulse, or "Rateau", after its French inventor, fr:Auguste Rateau. The Rateau employs simple impulse rotors separated by a nozzle diaphragm. The diaphragm is essentially a partition wall in the turbine with a series of tunnels cut into it, funnel shaped with the broad end facing the previous stage and the narrow the next they are also angled to direct the steam jets onto the impulse rotor.

- Mercury vapour turbines used mercury as the working fluid, to improve the efficiency of fossil-fuelled generating stations. Although a few power plants were built with combined mercury vapour and conventional steam turbines, the toxicity of the metal mercury was quickly apparent.

- Screw turbine is a water turbine which uses the principle of the Archimedean screw to convert the potential energy of water on an upstream level into kinetic energy.

Uses

Almost all electrical power on Earth is generated with a turbine of some type. Very high efficiency steam turbines harness around 40% of the thermal energy, with the rest exhausted as waste heat.

Most jet engines rely on turbines to supply mechanical work from their working fluid and fuel as do all nuclear ships and power plants.

Turbines are often part of a larger machine. A gas turbine, for example, may refer to an internal combustion machine that contains a turbine, ducts, compressor, combustor, heat-exchanger, fan and (in the case of one designed to produce electricity) an alternator. Combustion turbines and steam turbines may be connected to machinery such as pumps and compressors, or may be used for propulsion of ships, usually through an intermediate gearbox to reduce rotary speed.

Reciprocating piston engines such as aircraft engines can use a turbine powered by their exhaust to drive an intake-air compressor, a configuration known as a turbocharger (turbine supercharger) or, colloquially, a "turbo".

Turbines can have very high power density (i.e. the ratio of power to weight, or power to volume). This is because of their ability to operate at very high speeds. The Space Shuttle main engines used turbopumps (machines consisting of a pump driven by a turbine engine) to feed the propellants (liquid oxygen and liquid hydrogen) into the engine's combustion chamber. The liquid hydrogen turbopump is slightly larger than an automobile engine (weighing approximately 700 lb) and produces nearly 70,000 hp (52.2 MW).

Turboexpanders are widely used as sources of refrigeration in industrial processes.

Military jet engines, as a branch of gas turbines, have recently been used as primary flight controller in post-stall flight using jet deflections that are also called thrust vectoring. The U.S. Federal Aviation Administration has also conducted a study about civilizing such thrust vectoring systems to recover jetliners from catastrophes.

Turbine Map

Each turbine in a gas turbine engine has an operating map. Complete maps are either based on turbine rig test results or are predicted by a special computer program. Alternatively, the map of a similar turbine can be suitably scaled.

Description

A typical turbine map is shown on the right. In this particular case, lines of percent corrected speed (based on a reference value) are plotted against the x-axis which is pressure ratio, but deltaH/T (roughly proportional to temperature drop across the unit/component entry temperature) is also often used. The y-axis is some measure of flow, usually non-dimensional flow or, as in this case, corrected flow, but not actual flow. Sometimes the axes of a turbine map are transposed, to be consistent with those of a compressor map. As in this case, a companion plot, showing the variation of isentropic (i.e. adiabatic) or polytropic efficiency, is often also included.

In this example the turbine is a transonic unit, where the throat Mach number reaches sonic conditions and the turbine becomes truly choked. Consequently, there is virtually no variation in flow between the corrected speed lines at high pressure ratios.

Most turbines however, are subsonic devices, the highest Mach number at the NGV throat being about 0.85. Under these conditions, there is a slight scatter in flow between the percent corrected speed lines in the 'choked' region of the map, where the flow for a given speed reaches a plateau.

Unlike a compressor (or fan), surge (or stall) does not occur in a turbine. This is because the flow through the unit is all 'downhill', from high to low pressure. Consequently there is no surge line marked on a turbine map.

Working lines are difficult to see on a conventional turbine map because the speed lines bunch up. One trick is to replot the map, with the y-axis being the multiple of flow and corrected speed. This separates the speed lines, enabling working lines (and efficiency contours) to be cross-plotted and clearly seen.

Progressive Unchoking of the Expansion System

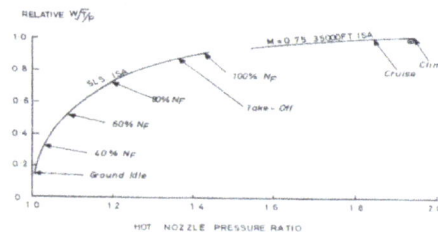

Typical primary nozzle map

The following discussion relates to the expansion system of a 2 spool, high bypass ratio, unmixed, turbofan.

On the RHS is a typical primary (i.e. hot) nozzle map (or characteristic). Its appearance is similar to that of a turbine map, but it lacks any (rotational) speed lines. Note that at high flight speeds (ignoring the change in altitude), the hot nozzle is usually in, or close to, a choking condition. This is because the ram rise in the air intake factors-up the nozzle pressure ratio. At static (e.g. SLS) conditions there is no ram rise, so the nozzle tends to operate unchoked (LHS of plot).

The low pressure turbine 'sees' the variation in flow capacity of the primary nozzle. A falling nozzle flow capacity tends to reduce the LP turbine pressure ratio (and deltaH/T). As the left hand map shows, initially the reduction in LP turbine deltaH/T has little effect upon the entry flow of the unit. Eventually, however, the LP turbine unchokes, causing the flow capacity of the LP turbine to start to decrease.

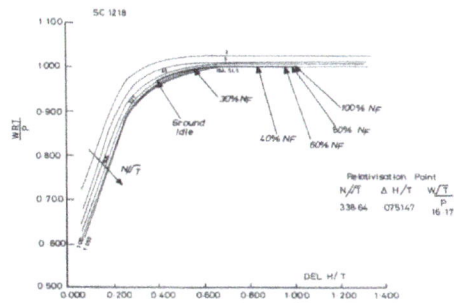

Typical high pressure turbine map

As long as the LP turbine remains choked, there is no significant change in HP turbine pressure ratio (or deltaH/T) and flow. Once, however, the LP turbine unchokes, the HP turbine deltaH/T starts to decrease. Eventually the HP turbine unchokes, causing its flow capacity to start to fall. Ground Idle is often reached shortly after HPT unchoke.

Steam Turbine

The rotor of a modern steam turbine used in a power plant

A steam turbine is a device that extracts thermal energy from pressurized steam and uses it to do mechanical work on a rotating output shaft. Its modern manifestation was invented by Sir Charles Parsons in 1884.

Because the turbine generates rotary motion, it is particularly suited to be used to drive an electrical generator – about 90% of all electricity generation in the United States (1996) is by use of steam turbines. The steam turbine is a form of heat engine that derives much of its improvement in thermodynamic efficiency from the use of multiple stages in the expansion of the steam, which results in a closer approach to the ideal reversible expansion process.

History

The first device that may be classified as a reaction steam turbine was little more than a toy, the classic Aeolipile, described in the 1st century by Greek mathematician Hero of Alexandria in Roman Egypt. In 1551, Taqi al-Din in Ottoman Egypt described a steam turbine with the practical application of rotating a spit. Steam turbines were also described by the Italian Giovanni Branca (1629) and John Wilkins in England (1648). The devices described by Taqi al-Din and Wilkins are today known as steam jacks. In 1672 an impulse steam turbine driven car was designed by Ferdinand Verbiest. A more modern version of this car was produced some time in the late 18th century by an unknown German mechanic.

A 250 kW industrial steam turbine from 1910 (right) directly linked to a generator (left).

The modern steam turbine was invented in 1884 by Sir Charles Parsons, whose first model was connected to a dynamo that generated 7.5 kW (10 hp) of electricity. The invention of Parsons' steam turbine made cheap and plentiful electricity possible and revolutionized marine transport and naval warfare. Parsons' design was a reaction type. His patent was licensed and the turbine scaled-up shortly after by an American, George Westinghouse. The Parsons turbine also turned out to be easy to scale up. Parsons had the satisfaction of seeing his invention adopted for all major world power stations, and the size of generators had increased from his first 7.5 kW set up to units of 50,000 kW capacity. Within Parson's lifetime, the generating capacity of a unit was scaled up by about 10,000 times, and the total output from turbo-generators constructed by his firm C. A. Parsons and Company and by their licensees, for land purposes alone, had exceeded thirty million horse-power.

A number of other variations of turbines have been developed that work effectively with steam. The *de Laval turbine* (invented by Gustaf de Laval) accelerated the steam to full speed before running it against a turbine blade. De Laval's impulse turbine is simpler, less expensive and does not need to be pressure-proof. It can operate with any pressure of steam, but is considerably less efficient. fr:Auguste Rateau developed a pressure compounded impulse turbine using the de Laval principle as early as 1896, obtained a US patent in 1903, and applied the turbine to a French torpedo boat in 1904. He taught at the École des mines de Saint-Étienne for a decade until 1897, and later founded a successful company that was incorporated into the Alstom firm after his death. One of the founders of the modern theory of steam and gas turbines was Aurel Stodola, a Slovak physicist and engineer and professor at the Swiss Polytechnical Institute (now ETH) in Zurich. His work *Die Dampfturbinen und ihre Aussichten als Wärmekraftmaschinen* (English: The Steam Turbine and its prospective use as a Heat Engine) was published in Berlin in 1903. A further book *Dampf und Gas-Turbinen* (English: Steam and Gas Turbines) was published in 1922.

The *Brown-Curtis turbine*, an impulse type, which had been originally developed and patented by the U.S. company International Curtis Marine Turbine Company, was developed in the 1900s in conjunction with John Brown & Company. It was used in John Brown-engined merchant ships and warships, including liners and Royal Navy warships.

Manufacturing

The present-day manufacturing industry for steam turbines is dominated by Chinese power equipment makers. Harbin Electric, Shanghai Electric, and Dongfang Electric, the top three power equipment makers in China, collectively hold a majority stake in the worldwide market share for steam turbines in 2009-10 according to Platts. Other manufacturers with minor market share include Bhel, Siemens, Alstom, GE, Mitsubishi Heavy Industries, and Toshiba. The consulting firm Frost & Sullivan projects that manufacturing of steam turbines will become more consolidated by 2020 as Chinese power manufacturers win increasing business outside of China.

Types

Steam turbines are made in a variety of sizes ranging from small <0.75 kW (<1 hp) units (rare) used as mechanical drives for pumps, compressors and other shaft driven equipment, to 1 500 000 kW (1.5 GW; 2 000 000 hp) turbines used to generate electricity. There are several classifications for modern steam turbines.

Blade and Stage Design

Turbine blades are of two basic types, blades and nozzles. Blades move entirely due to the impact of steam on them and their profiles do not converge. This results in a steam velocity drop and essentially no pressure drop as steam moves through the blades. A turbine composed of blades alternating with fixed nozzles is called an impulse turbine, Curtis turbine, Rateau turbine, or Brown-Curtis turbine. Nozzles appear similar to blades, but their profiles converge near the exit. This results in a steam pressure drop and velocity increase as steam moves through the nozzles. Nozzles move due to both the impact of steam on them and the reaction due to the high-velocity steam at the exit. A turbine composed of moving nozzles alternating with fixed nozzles is called a reaction turbine or Parsons turbine.

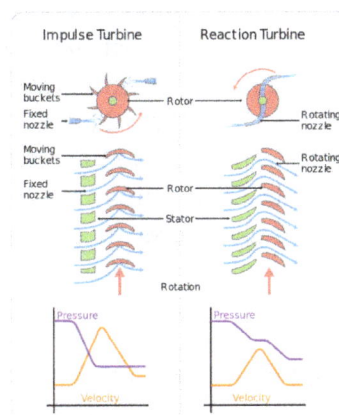

Schematic diagram outlining the difference between an impulse and a 50% reaction turbine

Except for low-power applications, turbine blades are arranged in multiple stages in series, called compounding, which greatly improves efficiency at low speeds. A reaction stage is a row of fixed nozzles followed by a row of moving nozzles. Multiple reaction stages divide the pressure drop between the steam inlet and exhaust into numerous small drops, resulting in a pressure-compounded turbine. Impulse stages may be either pressure-compounded, velocity-compounded, or pressure-velocity compounded. A pressure-compounded impulse stage is a row of fixed nozzles followed by a row of moving blades, with multiple stages for compounding. This is also known as a Rateau turbine, after its inventor. A velocity-compounded impulse stage (invented by Curtis and also called a "Curtis wheel") is a row of fixed nozzles followed by two or more rows of moving blades alternating with rows of fixed blades. This divides the velocity drop across the stage into several smaller drops. A series of velocity-compounded impulse stages is called a pressure-velocity compounded turbine.

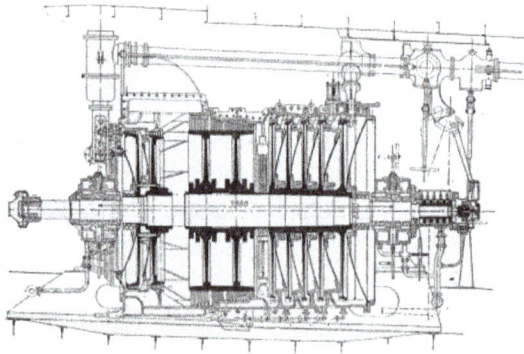

Diagram of an AEG marine steam turbine circa 1905

By 1905, when steam turbines were coming into use on fast ships (such as HMS *Dreadnought*) and in land-based power applications, it had been determined that it was desirable to use one or more Curtis wheels at the beginning of a multi-stage turbine (where the steam pressure is highest), followed by reaction stages. This was more efficient with high-pressure steam due to reduced leakage between the turbine rotor and the casing. This is illustrated in the drawing of the German 1905 AEG marine steam turbine. The steam from the boilers enters from the right at high pressure through a throttle, controlled manually by an operator (in this case a sailor known as the throttleman). It passes through five Curtis wheels and numerous reaction stages (the small blades at the edges of the two large rotors in the middle) before exiting at low pressure, almost certainly to a condenser. The condenser provides a vacuum that maximizes the energy extracted from the steam, and condenses the steam into feedwater to be returned to the boilers. On the left are several additional reaction stages (on two large rotors) that rotate the turbine in reverse for astern operation, with steam admitted by a separate throttle. Since ships are rarely operated in reverse, efficiency is not a priority in astern turbines, so only a few stages are used to save cost.

Blade Design Challenges

A major challenge facing turbine design is reducing the creep experienced by the blades. Because of the high temperatures and high stresses of operation, steam turbine materials become damaged through these mechanisms. As temperatures are increased in an effort to improve turbine efficiency, creep becomes more significant. To limit creep, thermal coatings and superalloys with solid-solution strengthening and grain boundary strengthening are used in blade designs.

Protective coatings are used to reduce the thermal damage and to limit oxidation. These coatings are often stabilized zirconium dioxide-based ceramics. Using a thermal protective coating limits the temperature exposure of the nickel superalloy. This reduces the creep mechanisms experienced in the blade. Oxidation coatings limit efficiency losses caused by a buildup on the outside of the blades, which is especially important in the high-temperature environment.

The nickel-based blades are alloyed with aluminum and titanium to improve strength and creep resistance. The microstructure of these alloys is composed of different regions of composition. A uniform dispersion of the gamma-prime phase – a combination of nickel, aluminum, and titanium – promotes the strength and creep resistance of the blade due to the microstructure.

Refractory elements such as rhenium and ruthenium can be added to the alloy to improve creep strength. The addition of these elements reduces the diffusion of the gamma prime phase, thus preserving the fatigue resistance, strength, and creep resistance.

Steam Supply and Exhaust Conditions

A low-pressure steam turbine in a nuclear power plant. These turbines exhaust steam at a pressure below atmospheric.

These types include condensing, non-condensing, reheat, extraction and induction.

Condensing turbines are most commonly found in electrical power plants. These turbines receive steam from a boiler and exhaust it to a condenser. The exhausted steam is at a pressure well below atmospheric, and is in a partially condensed state, typically of a quality near 90%.

Non-condensing or back pressure turbines are most widely used for process steam applications. The exhaust pressure is controlled by a regulating valve to suit the needs of the process steam pressure. These are commonly found at refineries, district heating units, pulp and paper plants, and desalination facilities where large amounts of low pressure process steam are needed.

Reheat turbines are also used almost exclusively in electrical power plants. In a reheat turbine, steam flow exits from a high pressure section of the turbine and is returned to the boiler where additional superheat is added. The steam then goes back into an intermediate pressure section of the turbine and continues its expansion. Using reheat in a cycle increases the work output from the turbine and also the expansion reaches conclusion before the steam condenses, thereby minimizing the erosion of the blades in last rows. In most of the cases, maximum number of reheats employed in a cycle is 2 as the cost of super-heating the steam negates the increase in the work output from turbine.

Extracting type turbines are common in all applications. In an extracting type turbine, steam is released from various stages of the turbine, and used for industrial process needs or sent to boiler feedwater heaters to improve overall cycle efficiency. Extraction flows may be controlled with a valve, or left uncontrolled.

Induction turbines introduce low pressure steam at an intermediate stage to produce additional power.

Casing or Shaft Arrangements

These arrangements include single casing, tandem compound and cross compound turbines. Single casing units are the most basic style where a single casing and shaft are coupled to a generator. Tandem compound are used where two or more casings are directly coupled together to drive a single generator. A cross compound turbine arrangement features two or more shafts not in line driving two or more generators that often operate at different speeds. A cross compound turbine is typically used for many large applications. A typical 1930s-1960s naval installation is illustrated below; this shows high- and low-pressure turbines driving a common reduction gear, with a geared cruising turbine on one high-pressure turbine.

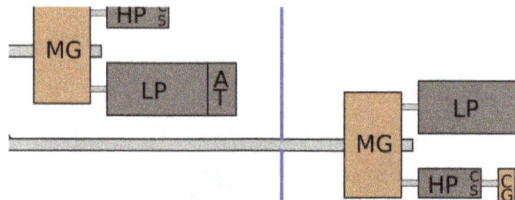

Starboard steam turbine machinery arrangement of Japanese *Furutaka*- and *Aoba*-class cruisers.

Two-Flow Rotors

A two-flow turbine rotor. The steam enters in the middle of the shaft, and exits at each end, balancing the axial force.

The moving steam imparts both a tangential and axial thrust on the turbine shaft, but the axial thrust in a simple turbine is unopposed. To maintain the correct rotor position and balancing, this force must be counteracted by an opposing force. Thrust bearings can be used for the shaft bearings, the rotor can use dummy pistons, it can be double flow- the steam enters in the middle of the shaft and exits at both ends, or a combination of any of these. In a double flow rotor, the blades in each half face opposite ways, so that the axial forces negate each other but the tangential forces act together. This design of rotor is also called two-flow, double-axial-flow, or double-exhaust. This arrangement is common in low-pressure casings of a compound turbine.

Principle of Operation and Design

An ideal steam turbine is considered to be an isentropic process, or constant entropy process, in which the entropy of the steam entering the turbine is equal to the entropy of the steam leaving the turbine. No steam turbine is truly isentropic, however, with typical isentropic efficiencies ranging from 20–90% based on the application of the turbine. The interior of a turbine comprises several sets of blades or *buckets*. One set of stationary blades is connected to the casing and one set of rotating blades is connected to the shaft. The sets intermesh with certain minimum clearances, with the size and configuration of sets varying to efficiently exploit the expansion of steam at each stage.

Theoretical Turbine Efficiency

To maximize turbine efficiency the steam is expanded, doing work, in a number of stages. These stages are characterized by how the energy is extracted from them and are known as either impulse or reaction turbines. Most steam turbines use a mixture of the reaction and impulse designs: each stage behaves as either one or the other, but the overall turbine uses both. Typically, lower pressure sections are reaction type and higher pressure stages are impulse type.

Impulse Turbines

An impulse turbine has fixed nozzles that orient the steam flow into high speed jets. These jets contain significant kinetic energy, which is converted into shaft rotation by the bucket-like shaped rotor blades, as the steam jet changes direction. A pressure drop occurs across only the stationary blades, with a net increase in steam velocity across the stage. As the steam flows through the nozzle its pressure falls from inlet pressure to the exit pressure (atmospheric pressure, or more usually, the condenser vacuum). Due to this high ratio of expansion of steam, the steam leaves the nozzle with a very high velocity. The steam leaving the moving blades has a large portion of the maximum velocity of the steam when leaving the nozzle. The loss of energy due to this higher exit velocity is commonly called the carry over velocity or leaving loss.

A selection of impulse turbine blades

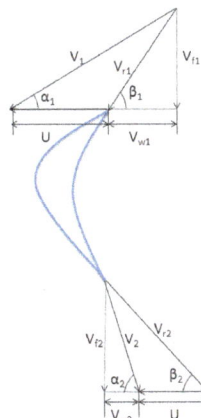

Velocity triangle

The law of moment of momentum states that the sum of the moments of external forces acting on a fluid which is temporarily occupying the control volume is equal to the net time change of angular momentum flux through the control volume.

The swirling fluid enters the control volume at radius r_1 with tangential velocity V_{w1} and leaves at radius V_{w1} with tangential velocity V_{w2}.

A velocity triangle paves the way for a better understanding of the relationship between the various velocities. In the adjacent figure we have:

V_1 and V_2 are the absolute velocities at the inlet and outlet respectively.

V_{f1} and V_{f2} are the flow velocities at the inlet and outlet respectively.

$V_{w1} + U$ and V_{w2} are the swirl velocities at the inlet and outlet respectively.

V_{r1} and V_{r2} are the relative velocities at the inlet and outlet respectively.

U_1 and U_2 are the velocities of the blade at the inlet and outlet respectively.

α is the guide vane angle and β is the blade angle.

Then by the law of moment of momentum, the torque on the fluid is given by:

$$T = \dot{m}(r_2 V_{w2} - r_1 V_{w1})$$

For an impulse steam turbine: $r_2 = r_1 = r$. Therefore, the tangential force on the blades is $F_u = \dot{m}(V_{w1} - V_{w2})$. The work done per unit time or power developed: $W = T * \omega$.

When ω is the angular velocity of the turbine, then the blade speed is $U = \omega * r$.. The power developed is then $W = \dot{m}U(\Delta V_w)$.

Blade Efficiency

Blade efficiency (η_b) can be defined as the ratio of the work done on the blades to kinetic energy supplied to the fluid, and is given by

$$\eta_b = \frac{Work\ Done}{Kinetic\ Energy\ Supplied} = \frac{2UV_w}{V_1^2}$$

Stage Efficiency

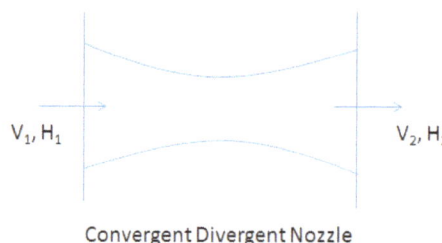

V_1, H_1　　　　　　　　　　　　　　　　V_2, H_2

Convergent Divergent Nozzle

Convergent-Divergent Nozzle

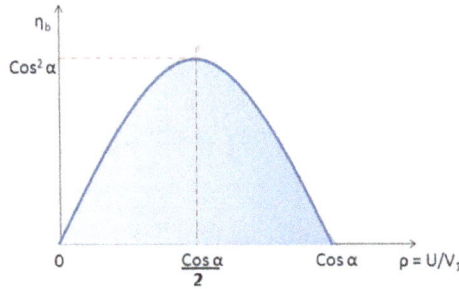

Graph depicting efficiency of Impulse turbine

A stage of an impulse turbine consists of a nozzle set and a moving wheel. The stage efficiency defines a relationship between enthalpy drop in the nozzle and work done in the stage.

$$\eta_{stage} = \frac{Work\ done\ on\ blade}{Energy\ supplied\ per\ stage} = \frac{U \Delta V_w}{\Delta h}$$

Where $\Delta h = h_2 - h_1$ is the specific enthalpy drop of steam in the nozzle.

By the first law of thermodynamics: $h_1 + \frac{V_1^2}{2} = h_2 + \frac{V_2^2}{2}$

Assuming that V_1 is appreciably less than V_2, we get $\Delta h \approx \frac{V_2^2}{2}$ Furthermore, stage efficiency is the

product of blade efficiency and nozzle efficiency, or $\eta_{stage} = \eta_b * \eta_N$

Nozzle efficiency is given by $\eta_N = \frac{V_2^2}{2(h_1 - h_2)}$, , where the enthalpy (in J/Kg) of steam at the en-

trance of the nozzle is $\Delta V_w = V_{w1} - (-V_{w2})$ and the enthalpy of steam at the exit of the nozzle is .

$$\Delta V_w = V_{w1} + V_{w2} \quad \Delta V_w = V_{r1} \cos \beta_1 + V_{r2} \cos \beta_2$$

$$\Delta V_w = V_{r1} \cos \beta_1 (1 + \frac{V_{r2} \cos \beta_2}{V_{r1} \cos \beta_1})$$

The ratio of the cosines of the blade angles at the outlet and inlet can be taken and denoted $c = \frac{\cos \beta_2}{\cos \beta_1}$. The ratio of steam velocities relative to the rotor speed at the outlet to the inlet of the blade is defined by the friction coefficient $k = \frac{V_{r2}}{V_{r1}}$.

$k < 1$ and and depicts the loss in the relative velocity due to friction as the steam flows around the blades ($k = 1$ for smooth blades).

$$\eta_b = \frac{2U \Delta V_w}{V_1^2} = \frac{2U(\cos \alpha_1 - U / V_1)(1 + kc)}{V_1}$$

The ratio of the blade speed to the absolute steam velocity at the inlet is termed the blade speed ratio $\rho = \frac{U}{V_1}$

η_b is maximum when $\dfrac{d\eta_b}{d\rho} = 0$ or, $\dfrac{d}{d\rho}(2\cos\alpha_1 - \rho^2(1+kc)) = 0$. That implies $\rho = \dfrac{\cos\alpha_1}{2}$ and there-

fore $\dfrac{U}{V_1} = \dfrac{\cos\alpha_1}{2}$.. Now $\rho_{opt} = \dfrac{U}{V_1} = \dfrac{\cos\alpha_1}{2}$ (for a single stage impulse turbine)

Therefore, the maximum value of stage efficiency is obtained by putting the value of $\dfrac{U}{V_1} = \dfrac{\cos\alpha_1}{2}$

in the expression of $\eta_b /$

We get: $(\eta_b)_{max} = 2(\rho\cos\alpha_1 - \rho^2)(1+kc) = \dfrac{\cos^2\alpha_1(1+kc)}{2}$.

For equiangular blades, $\beta_1 = \beta_2$, therefore $c = 1$, and we get $(\eta_b)_{max} = \dfrac{\cos^2\alpha_1(1+k)}{2}$. . If the friction due to the blade surface is neglected then $(\eta_b)_{max} = \cos^2\alpha_1$..

Conclusions on Maximum Efficiency

$$(\eta_b)_{max} = \cos^2\alpha_1$$

1. For a given steam velocity work done per kg of steam would be maximum when $\cos^2\alpha_1 = 1$ or $\alpha_1 = 0$.

2. As α_1 increases, the work done on the blades reduces, but at the same time surface area of the blade reduces, therefore there are less frictional losses.

Reaction turbines

In the *reaction turbine*, the rotor blades themselves are arranged to form convergent nozzles. This type of turbine makes use of the reaction force produced as the steam accelerates through the nozzles formed by the rotor. Steam is directed onto the rotor by the fixed vanes of the stator. It leaves the stator as a jet that fills the entire circumference of the rotor. The steam then changes direction and increases its speed relative to the speed of the blades. A pressure drop occurs across both the stator and the rotor, with steam accelerating through the stator and decelerating through the rotor, with no net change in steam velocity across the stage but with a decrease in both pressure and temperature, reflecting the work performed in the driving of the rotor.

Blade Efficiency

Energy input to the blades in a stage:

$E = \Delta h$ is equal to the kinetic energy supplied to the fixed blades (f) + the kinetic energy supplied to the moving blades (m).

Or, E = enthalpy drop over the fixed blades, Δh_f + enthalpy drop over the moving blades, Δh_m..

The effect of expansion of steam over the moving blades is to increase the relative velocity at the exit. Therefore, the relative velocity at the exit V_{r2} is always greater than the relative velocity at the inlet V_{r1}..

In terms of velocities, the enthalpy drop over the moving blades is given by:

$\Delta h_m = \dfrac{V_{r2}^2 - V_{r1}^2}{2}$ (it contributes to a change in static pressure)

The enthalpy drop in the fixed blades, with the assumption that the velocity of steam entering the fixed blades is equal to the velocity of steam leaving the previously moving blades is given by:

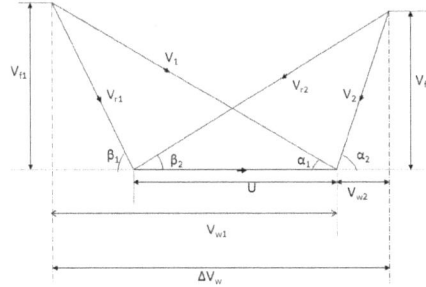

Velocity diagram

$\Delta h_f = \dfrac{V_1^2 - V_0^2}{2}$ where V_0 is the inlet velocity of steam in the nozzle

V_0 is very small and hence can be neglected

Therefore, $\Delta h_f = \dfrac{V_1^2}{2}$

$$E = \Delta h_f + \Delta h_m$$

$$E = \frac{V_1^2}{2} + \frac{V_{r2}^2 - V_{r1}^2}{2}$$

A very widely used design has half degree of reaction or 50% reaction and this is known as Parson's turbine. This consists of symmetrical rotor and stator blades. For this turbine the velocity triangle is similar and we have:

$$\alpha_1 = \beta_2, \ \beta_1 = \alpha_2$$

$$V_1 = V_{r2}, \ V_{r1} = V_2$$

Assuming *Parson's turbine* and obtaining all the expressions we get

$$E = V_1^2 - \frac{V_{r1}^2}{2}$$

From the inlet velocity triangle we have $V_{r1}^2 = V_1^2 + U^2 - 2UV_1 \cos\alpha_1$

$$E = V_1^2 - \frac{V_1^2}{2} - \frac{U^2}{2} + \frac{2UV_1 \cos\alpha_1}{2}$$

$$E = \frac{V_1^2 - U^2 + 2UV_1 \cos \alpha_1}{2}$$

Work done (for unit mass flow per second):

$$W = U * \Delta V_w = U * (2 * V_1 \cos \alpha_1 - U)$$

Therefore, the blade efficiency is given by

$$\eta_b = \frac{2U(2V_1 \cos \alpha_1 - U)}{V_1^2 - U^2 + 2V_1 U \cos \alpha_1}$$

Condition of Maximum Blade Efficiency

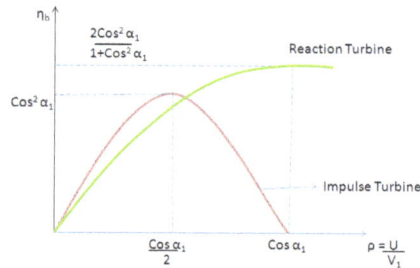

Comparing Efficiencies of Impulse and Reaction turbines

If $\rho = \dfrac{U}{V_1}$, then

$$(\eta_b)_{max} = \frac{2\rho(\cos \alpha_1 - \rho)}{V_1^2 - U^2 + 2UV_1 \cos \alpha_1}$$

For maximum efficiency $\dfrac{d\eta_b}{d\rho} = 0$, we get

$$(1 - \rho^2 + 2\rho \cos \alpha_1)(4 \cos \alpha_1 - 4\rho) - 2\rho(2 \cos \alpha_1 - \rho)(-2\rho + 2 \cos \alpha_1) = 0$$

and this finally gives $\rho_{opt} = \dfrac{U}{V_1} = \cos \alpha_1$

Therefore, $(\eta_b)_{max}$ is found by putting the value of $\rho = \cos \alpha_1$ in the expression of blade efficiency

$$(\eta_b)_{reaction} = \frac{2 \cos^2 \alpha_1}{1 + \cos^2 \alpha_1}$$

$$(\eta_b)_{impulse} = \cos^2 \alpha_1$$

Practical Turbine Efficiency

Practical thermal efficiency of a steam turbine varies with turbine size, load condition,gap losses and friction losses. They reach top values up to about 50% in a 1200MW turbine smaller ones have a lower efficiency.

Operation and Maintenance

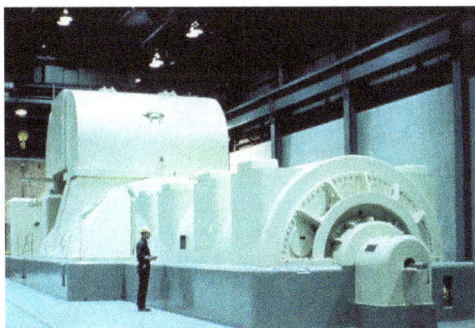

A modern steam turbine generator installation

Because of the high pressures used in the steam circuits and the materials used, steam turbines and their casings have high thermal inertia. When warming up a steam turbine for use, the main steam stop valves (after the boiler) have a bypass line to allow superheated steam to slowly bypass the valve and proceed to heat up the lines in the system along with the steam turbine. Also, a turning gear is engaged when there is no steam to slowly rotate the turbine to ensure even heating to prevent uneven expansion. After first rotating the turbine by the turning gear, allowing time for the rotor to assume a straight plane (no bowing), then the turning gear is disengaged and steam is admitted to the turbine, first to the astern blades then to the ahead blades slowly rotating the turbine at 10–15 RPM (0.17–0.25 Hz) to slowly warm the turbine. The warm up procedure for large steam turbines may exceed ten hours.

During normal operation, rotor imbalance can lead to vibration, which, because of the high rotation velocities, could lead to a blade breaking away from the rotor and through the casing. To reduce this risk, considerable efforts are spent to balance the turbine. Also, turbines are run with high quality steam: either superheated (dry) steam, or saturated steam with a high dryness fraction. This prevents the rapid impingement and erosion of the blades which occurs when condensed water is blasted onto the blades (moisture carry over). Also, liquid water entering the blades may damage the thrust bearings for the turbine shaft. To prevent this, along with controls and baffles in the boilers to ensure high quality steam, condensate drains are installed in the steam piping leading to the turbine.

Maintenance requirements of modern steam turbines are simple and incur low costs (typically around $0.005 per kWh); their operational life often exceeds 50 years.

Speed Regulation

The control of a turbine with a governor is essential, as turbines need to be run up slowly to prevent damage and some applications (such as the generation of alternating current electricity) require precise speed control. Uncontrolled acceleration of the turbine rotor can lead to an overspeed trip, which causes the governor and throttle valves that control the flow of steam to the turbine to close. If these valves fail then the turbine may continue accelerating until it breaks apart, often catastrophically. Turbines are expensive to make, requiring precision manufacture and special quality materials.

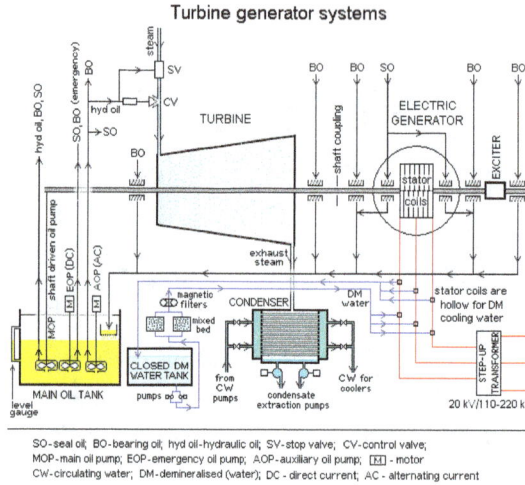

Diagram of a steam turbine generator system

During normal operation in synchronization with the electricity network, power plants are governed with a five percent droop speed control. This means the full load speed is 100% and the no-load speed is 105%. This is required for the stable operation of the network without hunting and drop-outs of power plants. Normally the changes in speed are minor. Adjustments in power output are made by slowly raising the droop curve by increasing the spring pressure on a centrifugal governor. Generally this is a basic system requirement for all power plants because the older and newer plants have to be compatible in response to the instantaneous changes in frequency without depending on outside communication.

Thermodynamics of Steam Turbines

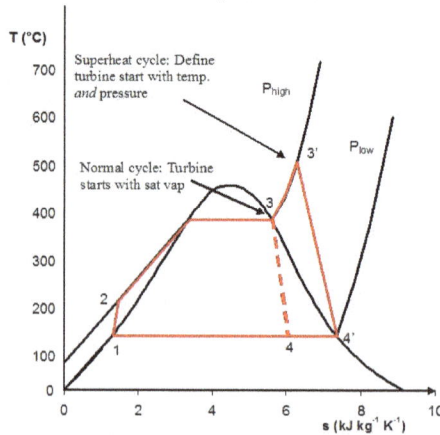

T-s diagram of a superheated Rankine cycle

The steam turbine operates on basic principles of thermodynamics using the part 3-4 of the Rankine cycle shown in the adjoining diagram. Superheated steam (or dry saturated steam, depending on application) leaves the boiler at high temperature and high pressure. At entry to the turbine, the steam gains kinetic energy by passing through a nozzle (a fixed nozzle in an impulse type turbine or the fixed blades in a reaction type turbine). When the steam leaves the nozzle it is moving at high velocity towards the blades of the turbine rotor. A force is created on the blades due to the pressure of the vapor on the blades causing them to move. A generator or other such device can be placed on the shaft, and the energy that was in the steam can now be stored and used. The steam leaves the turbine as a saturated vapor (or liquid-vapor mix depending on application) at a lower temperature and pressure than it entered with and is sent to the condenser to be cooled. The first law enables us to find an formula for the rate at which work is developed per unit mass. Assuming there is no heat transfer to the surrounding environment and that the changes in kinetic and potential energy are negligible compared to the change in specific enthalpy we arrive at the following equation

$$\frac{\dot{W}}{\dot{m}} = h_3 - h_4$$

where

- is the rate at which work is developed per unit time
- is the rate of mass flow through the turbine

Isentropic Efficiency

To measure how well a turbine is performing we can look at its isentropic efficiency. This compares the actual performance of the turbine with the performance that would be achieved by an ideal, isentropic, turbine. When calculating this efficiency, heat lost to the surroundings is assumed to be zero. The starting pressure and temperature is the same for both the actual and the ideal turbines, but at turbine exit the energy content ('specific enthalpy') for the actual turbine is greater than that for the ideal turbine because of irreversibility in the actual turbine. The specific enthalpy is evaluated at the same pressure for the actual and ideal turbines in order to give a good comparison between the two.

The isentropic efficiency is found by dividing the actual work by the ideal work.

$$\eta_t = \frac{h_3 - h_4}{h_3 - h_{4s}}$$

where

- h_3 is the specific enthalpy at state three
- h_4 is the specific enthalpy at state 4 for the actual turbine
- h_{4s} is the specific enthalpy at state 4s for the isentropic turbine

(but note that the adjacent diagram does not show state 4s: it is vertically below state 3)

Direct Drive

A direct-drive 5 MW steam turbine fuelled with biomass

Electrical power stations use large steam turbines driving electric generators to produce most (about 80%) of the world's electricity. The advent of large steam turbines made central-station electricity generation practical, since reciprocating steam engines of large rating became very bulky, and operated at slow speeds. Most central stations are fossil fuel power plants and nuclear power plants; some installations use geothermal steam, or use concentrated solar power (CSP) to create the steam. Steam turbines can also be used directly to drive large centrifugal pumps, such as feedwater pumps at a thermal power plant.

The turbines used for electric power generation are most often directly coupled to their generators. As the generators must rotate at constant synchronous speeds according to the frequency of the electric power system, the most common speeds are 3,000 RPM for 50 Hz systems, and 3,600 RPM for 60 Hz systems. Since nuclear reactors have lower temperature limits than fossil-fired plants, with lower steam quality, the turbine generator sets may be arranged to operate at half these speeds, but with four-pole generators, to reduce erosion of turbine blades.

Marine Propulsion

Turbinia, 1894, the first steam turbine-powered ship

In steamships, advantages of steam turbines over reciprocating engines are smaller size, lower maintenance, lighter weight, and lower vibration. A steam turbine is only efficient when operating in the thousands of RPM, while the most effective propeller designs are for speeds less than 300 RPM; consequently, precise (thus expensive) reduction gears are usually required, although numerous early ships through World War I, such as *Turbinia*, had direct drive from the steam turbines to the propeller shafts. Another alternative is turbo-electric transmission, in which an electrical generator run by the high-speed turbine is used to run one or more slow-speed electric motors connected to the propeller shafts; precision gear cutting may be a production bottleneck during wartime. Turbo-electric drive was most used in large US warships designed during World War I and in some fast liners, and was used in some troop transports and mass-production destroyer escorts in World War II.

High and low pressure turbines for SS *Maui*.

The higher cost of turbines and the associated gears or generator/motor sets is offset by lower maintenance requirements and the smaller size of a turbine when compared to a reciprocating engine having an equivalent power, although the fuel costs are higher than a diesel engine because steam turbines have lower thermal efficiency. To reduce fuel costs the thermal efficiency of both types of engine have been improved over the years. Today, propulsion steam turbine cycle efficiencies have yet to break 50%, yet diesel engines routinely exceed 50%, especially in marine applications. Diesel power plants also have lower operating costs since fewer operators are required. Thus, conventional steam power is used in very few new ships. An exception is LNG carriers which often find it more economical to use boil-off gas with a steam turbine than to re-liquify it.

Parsons turbine from the 1928 Polish destroyer *Wicher*.

Nuclear-powered ships and submarines use a nuclear reactor to create steam for turbines. Nuclear power is often chosen where diesel power would be impractical (as in submarine applications) or the logistics of refuelling pose significant problems (for example, icebreakers). It has been estimated that the reactor fuel for the Royal Navy's *Vanguard*-class submarines is sufficient to last 40 circumnavigations of the globe – potentially sufficient for the vessel's entire service life. Nuclear propulsion has only been applied to a very few commercial vessels due to the expense of maintenance and the regulatory controls required on nuclear systems and fuel cycles.

Early Development

The development of steam turbine marine propulsion from 1894-1935 was dominated by the need to reconcile the high efficient speed of the turbine with the low efficient speed (less than 300 rpm) of the ship's propeller at an overall cost competitive with reciprocating engines. In 1894, efficient reduction gears were not available for the high powers required by ships, so direct drive was necessary. In *Turbinia*, which has direct drive to each propeller shaft, the efficient speed of the turbine was reduced after initial trials by directing the steam flow through all three direct drive turbines (one on each shaft) in series, probably totaling around 200 turbine stages operating in series. Also, there were three propellers on each shaft for operation at high speeds. The high shaft speeds of the era are represented by one of the first US turbine-powered destroyers, USS *Smith*, launched in 1909, which had direct drive turbines and whose three shafts turned at 724 rpm at 28.35 knots. The use of turbines in several casings exhausting steam to each other in series became standard in most subsequent marine propulsion applications, and is a form of cross-compounding. The first turbine was called the high pressure (HP) turbine, the last turbine was the low pressure (LP) turbine, and any turbine in between was an intermediate pressure (IP) turbine. A much later arrangement than *Turbinia* can be seen on RMS *Queen Mary* in Long Beach, California, launched in 1934, in which each shaft is powered by four turbines in series connected to the ends of the two input shafts of a single-reduction gearbox. They are the HP, 1st IP, 2nd IP, and LP turbines.

Cruising Machinery and Gearing

The quest for economy was even more important when cruising speeds were considered. Cruising speed is roughly 50% of a warship's maximum speed and 20-25% of its maximum power level. This would be a speed used on long voyages when fuel economy is desired. Although this brought the propeller speeds down to an efficient range, turbine efficiency was greatly reduced, and early turbine ships had poor cruising ranges. A solution that proved useful through most of the steam turbine propulsion era was the cruising turbine. This was an extra turbine to add even more stages, at first attached directly to one or more shafts, exhausting to a stage partway along the HP turbine, and not used at high speeds. As reduction gears became available around 1911, some ships, notably the battleship USS *Nevada*, had them on cruising turbines while retaining direct drive main turbines. Reduction gears allowed turbines to operate in their efficient range at a much higher speed than the shaft, but were expensive to manufacture.

Cruising turbines competed at first with reciprocating engines for fuel economy. An example of the retention of reciprocating engines on fast ships was the famous RMS *Titanic* of 1911, which along with her sisters RMS *Olympic* and HMHS *Britannic* had triple-expansion engines on the two outboard shafts, both exhausting to an LP turbine on the center shaft. After adopting turbines with the *Delaware*-class battleships launched in 1909, the United States Navy reverted to reciprocating machinery on the *New York*-class battleships of 1912, then went back to turbines on *Nevada* in 1914. The lingering fondness for reciprocating machinery was because the US Navy had no plans for capital ships exceeding 21 knots until after World War I, so top speed was less important than economical cruising. The United States had acquired the Philippines and Hawaii as territories in 1898, and lacked the British Royal Navy's worldwide network of coaling stations. Thus, the US Navy in 1900-1940 had the greatest need of any nation for fuel economy, especially as the prospect of war with Japan arose following World War I. This need was compounded by the US not launching any cruisers 1908-1920, so destroyers were required to perform long-range missions usually assigned to cruisers. So, various cruising solutions were fitted on US destroyers launched 1908-1916. These included small reciprocating engines and geared or ungeared cruising turbines on one or two shafts. However, once fully geared turbines proved economical in initial cost and fuel they were rapidly adopted, with cruising turbines also included on most ships. Beginning in 1915 all new Royal Navy destroyers had fully geared turbines, and the United States followed in 1917.

In the Royal Navy, speed was a priority until the Battle of Jutland in mid-1916 showed that in the battlecruisers too much armour had been sacrificed in its pursuit. The British used exclusively turbine-powered warships from 1906. Because they recognized that a significant cruising range would be desirable given their world-wide empire, some warships, notably the *Queen Elizabeth*-class battleships, were fitted with cruising turbines from 1912 onwards following earlier experimental installations.

In the US Navy, the *Mahan*-class destroyers, launched 1935-36, introduced double-reduction gearing. This further increased the turbine speed above the shaft speed, allowing smaller turbines than single-reduction gearing. Steam pressures and temperatures were also increasing progressively, from 300 psi/425 F (2.07 MPa/218 C)(saturation temperature) on the World War I-era *Wickes* class to 615 psi/850 F (4.25 MPa/454 C) superheated steam on some World War II *Fletcher*-class destroyers and later ships. A standard configuration emerged of an axial-flow high pressure turbine (sometimes with a cruising turbine attached) and a double-axial-flow low pressure turbine connected to a double-reduction gearbox. This arrangement continued throughout the steam era in the US Navy and was also used in some Royal Navy designs. Machinery of this configuration can be seen on many preserved World War II-era warships in several countries. When US Navy warship construction resumed in the early 1950s, most surface combatants and aircraft carriers used 1,200 psi/950 F (8.28 MPa/510 C) steam. This continued until the end of the US Navy steam-powered warship era with the *Knox*-class frigates of the early 1970s. Amphibious and auxiliary ships continued to use 600 psi (4.14 MPa) steam post-World War II, with USS *Iwo Jima*, launched in 2001, possibly the last non-nuclear steam-powered ship built for the US Navy.

Turbo-Electric Drive

NS *50 Let Pobedy*, a nuclear icebreaker with nuclear-turbo-electric propulsion

Turbo-electric drive was introduced on the battleship USS *New Mexico*, launched in 1917. Over the next eight years the US Navy launched five additional turbo-electric-powered battleships and two aircraft carriers (initially ordered as *Lexington*-class battlecruisers). Ten more turbo-electric capital ships were planned, but cancelled due to the limits imposed by the Washington Naval Treaty. Although *New Mexico* was refitted with geared turbines in a 1931-33 refit, the remaining turbo-electric ships retained the system throughout their careers. This system used two large steam turbine generators to drive an electric motor on each of four shafts. The system was less costly initially than reduction gears and made the ships more maneuverable in port, with the shafts able to reverse rapidly and deliver more reverse power than with most geared systems. Some ocean liners were also built with turbo-electric drive, as were some troop transports and mass-production destroyer escorts in World War II. However, when the US designed the "treaty cruisers", beginning with USS *Pensacola* launched in 1927, geared turbines were used to conserve weight, and remained in use for all fast steam-powered ships thereafter.

Current Usage

Since the 1980s, steam turbines have been replaced by gas turbines on fast ships and by diesel engines on other ships; exceptions are nuclear-powered ships and submarines and LNG carriers. Some auxiliary ships continue to use steam propulsion. In the U.S. Navy, the conventionally powered steam turbine is still in use on all but one of the Wasp-class amphibious assault ships. The U.S. Navy also operates steam turbines on their nuclear powered Nimitz-class and Ford-class aircraft carriers along with all of their nuclear submarines (Ohio-, Los Angeles-, Seawolf-, and Virginia-classes). The Royal Navy decommissioned its last conventional steam-powered surface warship class, the *Fearless*-class landing platform dock, in 2002. In 2013, the French Navy ended its steam era with the decommissioning of its last *Tourville*-class frigate. Amongst the other blue-water navies, the Russian Navy currently operates steam-powered *Kuznetsov*-class aircraft carriers and *Sovremenny*-class destroyers. The Indian Navy currently operates two conventional steam-powered carriers, INS *Viraat*, a former British *Centaur*-class aircraft carrier (to be decommissioned in 2016), and INS *Vikramaditya*, a modified *Kiev*-class aircraft carrier; it also operates three *Brahmaputra*-class frigates commissioned in the early 2000s and two *Godavari*-class frigates currently scheduled for decommissioning.

Most other naval forces either retired or re-engined their steam-powered warships by 2010. The Chinese Navy currently operates steam-powered Russian *Kuznetsov*-class aircraft carriers and *Sovremenny*-class destroyers; it also operates steam-powered *Luda*-class destroyers. The JS *Kurama*, the last steam-powered JMSDF *Shirane*-class destroyer, will be decommissioned and replaced in 2017. as of 2016, the Brazilian Navy operates *São Paulo*, a former French *Clemenceau*-class aircraft carrier, while the Mexican Navy currently operates four former U.S. *Knox*-class frigates and two former U.S. *Bronstein*-class frigates. The Royal Thai Navy, Egyptian Navy and the Republic of China Navy respectively operate one, two and six former U.S. *Knox*-class frigates. The Peruvian Navy currently operates the former Dutch *De Zeven Provinciën*-class cruiser *BAP Almirante Grau*; the Ecuadorian Navy currently operates two *Condell*-class frigates (modified *Leander*-class frigates).

Locomotives

A steam turbine locomotive engine is a steam locomotive driven by a steam turbine.

The main advantages of a steam turbine locomotive are better rotational balance and reduced hammer blow on the track. However, a disadvantage is less flexible output power so that turbine locomotives were best suited for long-haul operations at a constant output power.

The first steam turbine rail locomotive was built in 1908 for the Officine Meccaniche Miani Silvestri Grodona Comi, Milan, Italy. In 1924 Krupp built the steam turbine locomotive T18 001, operational in 1929, for Deutsche Reichsbahn.

Testing

British, German, other national and international test codes are used to standardize the procedures and definitions used to test steam turbines. Selection of the test code to be used is an agreement between the purchaser and the manufacturer, and has some significance to the design of the turbine and associated systems. In the United States, ASME has produced several performance test codes on steam turbines. These include ASME PTC 6-2004, Steam Turbines,

ASME PTC 6.2-2011, Steam Turbines in Combined Cycles, PTC 6S-1988, Procedures for Routine Performance Test of Steam Turbines. These ASME performance test codes have gained international recognition and acceptance for testing steam turbines. The single most important and differentiating characteristic of ASME performance test codes, including PTC 6, is that the test uncertainty of the measurement indicates the quality of the test and is not to be used as a commercial tolerance.

Stator

Rotor (lower left) and stator (upper right) of an electric motor

Stator of a 3-phase AC-motor

Stator of a brushless DC motor from computer cooler fan.

The stator is the stationary part of a rotary system, found in electric generators, electric motors, sirens, or biological rotors. The main use of a stator is to keep the field aligned.

In Motors

Depending on the configuration of a spinning electromotive device the stator may act as the *field magnet*, interacting with the armature to create motion, or it may act as the *armature*, receiving its influence from moving field coils on the rotor. The first DC generators (known as dynamos) and DC motors put the field coils on the stator, and the power generation or motive reaction coils on the rotor. This is necessary because a continuously moving power switch known as the commutator is needed to keep the field correctly aligned across the spinning rotor. The commutator must become larger and more robust as the current increases.

Stator winding of a generator at a hydroelectric power station.

The stator of these devices may be either a permanent magnet or an electromagnet. Where the stator is an electromagnet, the coil which energizes it is known as the *field coil* or *field winding*.

The coil can be either iron core or aluminum. To reduce loading losses in motors, manufacturers invariably use copper as the conducting material in windings. Aluminum, because of its lower electrical conductivity, may be an alternate material in fractional horsepower motors, especially when the motors are used for very short durations.

An AC alternator is able to produce power across multiple high-current power generation coils connected in parallel, eliminating the need for the commutator. Placing the field coils on the rotor allows for an inexpensive slip ring mechanism to transfer high-voltage, low current power to the rotating field coil.

It consists of a steel frame enclosing a hollow cylindrical core (made up of laminations of silicon steel). The laminations are to reduce hysteresis and eddy current losses.

Fluid Devices

In a turbine, the stator element contains blades or ports used to redirect the flow of fluid. Such devices include the steam turbine and the torque converter. In a mechanical siren, the stator contains one or more rows of holes that admit air into the rotor; by controlling the flow of air through the holes, the sound of the siren can be altered.

References

- Wiser, Wendell H. (2000). Energy resources: occurrence, production, conversion, use. Birkhäuser. p. 190. ISBN 978-0-387-98744-6.

- Whitaker, Jerry C. (2006). AC power systems handbook. Boca Raton, FL: Taylor and Francis. p. 35. ISBN 978-0-8493-4034-5.

- Leyzerovich, Alexander (2005). Wet-steam Turbines for Nuclear Power Plants. Tulsa OK: PennWell Books. p. 111. ISBN 978-1-59370-032-4.

- William P. Sanders (ed), Turbine Steam Path Mechanical Design and Manufacture, Volume Iiia (PennWell Books, 2004) ISBN 1-59370-009-1 page 292

- Energy and Environmental Analysis (2008). "Technology Characterization: Steam Turbines (2008)" (PDF). Report prepared for U.S. Environmental Protection Agency. p. 13. Retrieved 25 February 2013.

- "New Benchmarks for Steam Turbine Efficiency - Power Engineering". Pepei.pennnet.com. Archived from the original on 2010-11-18. Retrieved 2010-09-12.

- Encyclopædia Britannica (1931-02-11). "Sir Charles Algernon Parsons (British engineer) - Britannica Online Encyclopedia". Britannica.com. Retrieved 2010-09-12.

2

Components of a Gas Turbine

The major components of a gas turbine are a gas compressor and a combustion chamber. The gas compressor pumps in atmospheric air and brings it to high pressure while fuel is sprayed and ignited in the combustion chamber to produce energy. The chapter details the whole process that converts fuel into energy with detailed step by step information about the functioning of a gas turbine. This chapter is an overview of the subject matter incorporating all the major aspects of gas turbines.

Gas Compressor

A small stationary high pressure breathing air compressor for filling scuba cylinders

A gas compressor is a mechanical device that increases the pressure of a gas by reducing its volume. An air compressor is a specific type of gas compressor.

Compressors are similar to pumps: both increase the pressure on a fluid and both can transport the fluid through a pipe. As gases are compressible, the compressor also reduces the volume of a gas. Liquids are relatively incompressible; while some can be compressed, the main action of a pump is to pressurize and transport liquids.

Types of Compressors

The main types of gas compressors are illustrated and discussed below:

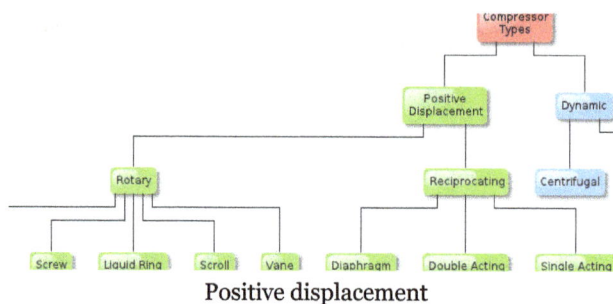

Positive displacement

Reciprocating Compressors

A motor-driven six-cylinder reciprocating compressor that can operate with two, four or six cylinders.

Reciprocating compressors use pistons driven by a crankshaft. They can be either stationary or portable, can be single or multi-staged, and can be driven by electric motors or internal combustion engines. Small reciprocating compressors from 5 to 30 horsepower (hp) are commonly seen in automotive applications and are typically for intermittent duty. Larger reciprocating compressors well over 1,000 hp (750 kW) are commonly found in large industrial and petroleum applications. Discharge pressures can range from low pressure to very high pressure (>18000 psi or 180 MPa). In certain applications, such as air compression, multi-stage double-acting compressors are said to be the most efficient compressors available, and are typically larger, and more costly than comparable rotary units. Another type of reciprocating compressor is the swash plate compressor, which uses pistons moved by a swash plate mounted on a shaft.

Household, home workshop, and smaller job site compressors are typically reciprocating compressors 1½ hp or less with an attached receiver tank.

Ionic Liquid Piston Compressor

An ionic liquid piston compressor, *ionic compressor* or *ionic liquid piston pump* is a hydrogen compressor based on an ionic liquid piston instead of a metal piston as in a piston-metal diaphragm compressor.

Rotary Screw Compressors

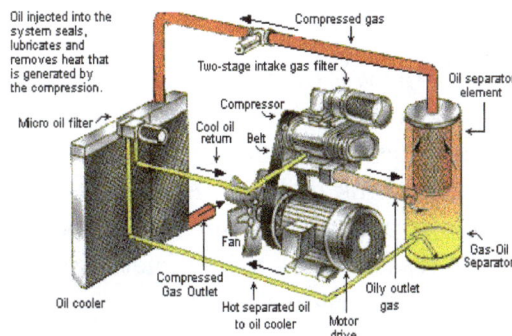

Diagram of a rotary screw compressor

Rotary screw compressors use two meshed rotating positive-displacement helical screws to force the gas into a smaller space. These are usually used for continuous operation in commercial and industrial applications and may be either stationary or portable. Their application can be from 3 horsepower (2.2 kW) to over 1,200 horsepower (890 kW) and from low pressure to moderately high pressure (>1,200 psi or 8.3 MPa).

Rotary screw compressors are commercially produced in Oil Flooded, Water Flooded and Dry type. The efficiency of rotary compressors depends on the air drier, and the selection of air drier is always 1.5 times volumetric delivery of the compressor.

Rotary Vane Compressors

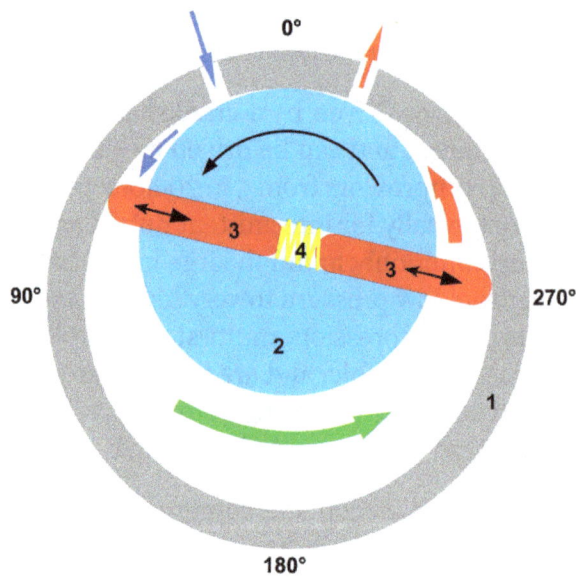

Eccentric rotary-vane pump

Rotary vane compressors consist of a rotor with a number of blades inserted in radial slots in the rotor. The rotor is mounted offset in a larger housing that is either circular or a more complex shape. As the rotor turns, blades slide in and out of the slots keeping contact with the outer wall of the housing. Thus, a series of increasing and decreasing volumes is created by the rotating blades. Rotary Vane compressors are, with piston compressors one of the oldest of compressor technologies.

With suitable port connections, the devices may be either a compressor or a vacuum pump. They can be either stationary or portable, can be single or multi-staged, and can be driven by electric motors or internal combustion engines. Dry vane machines are used at relatively low pressures (e.g., 2 bar or 200 kPa or 29 psi) for bulk material movement while oil-injected machines have the necessary volumetric efficiency to achieve pressures up to about 13 bar (1,300 kPa; 190 psi) in a single stage. A rotary vane compressor is well suited to electric motor drive and is significantly quieter in operation than the equivalent piston compressor.

Rotary vane compressors can have mechanical efficiencies of about 90%.

Rolling Piston

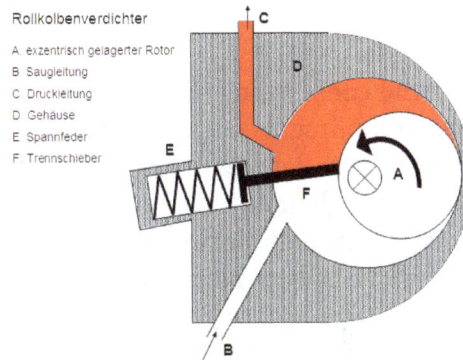

Rolling piston compressor

Rolling piston forces gas against a stationary vane.

Scroll Compressors

Mechanism of a scroll pump

A scroll compressor, also known as scroll pump and scroll vacuum pump, uses two interleaved spiral-like vanes to pump or compress fluids such as liquids and gases. The vane geometry may be involute, archimedean spiral, or hybrid curves. They operate more smoothly, quietly, and reliably than other types of compressors in the lower volume range.

Often, one of the scrolls is fixed, while the other orbits eccentrically without rotating, thereby trapping and pumping or compressing pockets of fluid between the scrolls.

Due to minimum clearance volume between the fixed scroll and the orbiting scroll, these compressors have a very high volumetric efficiency.

This type of compressor was used as the supercharger on Volkswagen G60 and G40 engines in the early 1990s.

Diaphragm Compressors

A diaphragm compressor (also known as a membrane compressor) is a variant of the conventional reciprocating compressor. The compression of gas occurs by the movement of a flexible membrane, instead of an intake element. The back and forth movement of the membrane is driven by a rod and a crankshaft mechanism. Only the membrane and the compressor box come in contact with the gas being compressed.

The degree of flexing and the material constituting the diaphragm affects the maintenance life of the equipment. Generally stiff metal diaphragms may only displace a few cubic centimeters of volume because the metal can not endure large degrees of flexing without cracking, but the stiffness of a metal diaphragm allows it to pump at high pressures. Rubber or silicone diaphragms are capable of enduring deep pumping strokes of very high flexion, but their low strength limits their use to low-pressure applications, and they need to be replaced as plastic embrittlement occurs.

Diaphragm compressors are used for hydrogen and compressed natural gas (CNG) as well as in a number of other applications.

A three-stage diaphragm compressor

The photograph on the right depicts a three-stage diaphragm compressor used to compress hydrogen gas to 6,000 psi (41 MPa) for use in a prototype compressed hydrogen and compressed natural gas (CNG) fueling station built in downtown Phoenix, Arizona by the Arizona Public Service company (an electric utilities company). Reciprocating compressors were used to compress the natural gas. The reciprocating natural gas compressor was developed by Sertco.

The prototype alternative fueling station was built in compliance with all of the prevailing safety, environmental and building codes in Phoenix to demonstrate that such fueling stations could be built in urban areas.

Dynamic

Dynamic compressors depend upon the inertia and momentum of a fluid.

Air Bubble Compressor

Also known as a trompe. A mixture of air and water generated through turbulence is allowed to fall into a subterranean chamber where the air separates from the water. The weight of falling water compresses the air in the top of the chamber. A submerged outlet from the chamber allows water to flow to the surface at a lower height than the intake. An outlet in the roof of the chamber supplies the compressed air to the surface. A facility on this principle was built on the Montreal River at Ragged Shutes near Cobalt, Ontario in 1910 and supplied 5,000 horsepower to nearby mines.

Centrifugal Compressors

A single stage centrifugal compressor

Centrifugal compressors use a rotating disk or impeller in a shaped housing to force the gas to the rim of the impeller, increasing the velocity of the gas. A diffuser (divergent duct) section converts the velocity energy to pressure energy. They are primarily used for continuous, stationary service in industries such as oil refineries, chemical and petrochemical plants and natural gas processing plants. Their application can be from 100 horsepower (75 kW) to thousands of horsepower. With multiple staging, they can achieve high output pressures greater than 10,000 psi (69 MPa).

Many large snowmaking operations (like ski resorts) use this type of compressor. They are also used in internal combustion engines as superchargers and turbochargers. Centrifugal compressors are used in small gas turbine engines or as the final compression stage of medium-sized gas turbines.

Diagonal or Mixed-Flow Compressors

Diagonal or mixed-flow compressors are similar to centrifugal compressors, but have a radial and axial velocity component at the exit from the rotor. The diffuser is often used to turn diagonal flow to an axial rather than radial direction.

Axial-Flow Compressors

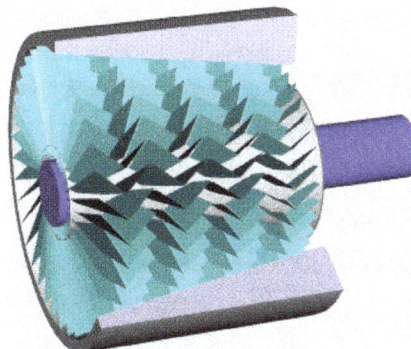

An animation of an axial compressor.

Axial-flow compressors are dynamic rotating compressors that use arrays of fan-like airfoils to progressively compress a fluid. They are used where high flow rates or a compact design are required.

The arrays of airfoils are set in rows, usually as pairs: one rotating and one stationary. The rotating airfoils, also known as blades or *rotors*, accelerate the fluid. The stationary airfoils, also known as *stators* or vanes, decelerate and redirect the flow direction of the fluid, preparing it for the rotor blades of the next stage. Axial compressors are almost always multi-staged, with the cross-sectional area of the gas passage diminishing along the compressor to maintain an optimum axial Mach number. Beyond about 5 stages or a 4:1 design pressure ratio a compressor will not function unless fitted with features such as stationary vanes with variable angles (known as variable inlet guide vanes and variable stators), the ability to allow some air to escape part-way along the compressor (known as interstage bleed) and being split into more than one rotating assembly (known as twin spools, for example).

Axial compressors can have high efficiencies; around 90% polytropic at their design conditions. However, they are relatively expensive, requiring a large number of components, tight tolerances and high quality materials. Axial-flow compressors are used in medium to large gas turbine engines, natural gas pumping stations, and some chemical plants.

Hermetically Sealed, Open, or Semi-Hermetic

A small hermetically sealed compressor in a common consumer refrigerator or freezer typically has a rounded steel outer shell permanently welded shut, which seals operating gases inside the system. There is no route for gases to leak, such as around motor shaft seals. On this model, the plastic top section is part of an auto-defrost system that uses motor heat to evaporate the water.

Compressors used in refrigeration systems are often described as being either hermetic, open, or semi-hermetic, to describe how the compressor and motor drive are situated in relation to the gas or vapor being compressed. The industry name for a hermetic is hermetically sealed compressor, while a semi-hermetic is commonly called a semi-hermetic compressor.

In hermetic and most semi-hermetic compressors, the compressor and motor driving the compressor are integrated, and operate within the pressurized gas envelope of the system. The motor is designed to operate in, and be cooled by, the refrigerant gas being compressed.

The difference between the hermetic and semi-hermetic, is that the hermetic uses a one-piece welded steel casing that cannot be opened for repair; if the hermetic fails it is simply replaced with an entire new unit. A semi-hermetic uses a large cast metal shell with gasketed covers that can be opened to replace motor and pump components.

The primary advantage of a hermetic and semi-hermetic is that there is no route for the gas to leak out of the system. Open compressors rely on shaft seals to retain the internal pressure, and these seals require a lubricant such as oil to retain their sealing properties.

An open pressurized system such as an automobile air conditioner can be more susceptible to leak its operating gases. Open systems rely on lubricant in the system to splash on pump components and seals. If it is not operated frequently enough, the lubricant on the seals slowly evaporates, and then the seals begin to leak until the system is no longer functional and must be recharged. By comparison, a hermetic system can sit unused for years, and can usually be started up again at any time without requiring maintenance or experiencing any loss of system pressure.

The disadvantage of hermetic compressors is that the motor drive cannot be repaired or maintained, and the entire compressor must be replaced if a motor fails. A further disadvantage is that burnt-out windings can contaminate whole systems, thereby requiring the system to be entirely pumped down and the gas replaced. Typically, hermetic compressors are used in low-cost factory-assembled consumer goods where the cost of repair is high compared to the value of the device, and it would be more economical to just purchase a new device.

An advantage of open compressors is that they can be driven by non-electric power sources, such as an internal combustion engine or turbine. However, open compressors that drive refrigeration systems are generally not totally *maintenance-free* throughout the life of the system, since some gas leakage will occur over time.

Thermodynamics of Gas Compression

Isentropic Compressor

A compressor can be idealized as internally reversible and adiabatic, thus an isentropic steady state device, meaning the change in entropy is 0. By defining the compression cycle as isentropic, an ideal efficiency for the process can be attained, and the ideal compressor performance can be compared to the actual performance of the machine. Isotropic Compression as used in ASME PTC 10 Code refers to a reversible, adiabatic compression process

Isentropic efficiency of Compressors:

$$\eta_C = \frac{\text{Isentropic Compressor Work}}{\text{Actual Compressor Work}} = \frac{W_s}{W_a} \cong \frac{h_{2s} - h_1}{h_{2a} - h_1}$$

h_1 is the enthalpy at the initial state

h_{2a} is the enthalpy at the final state for the actual process

h_{2s} is the enthalpy at the final state for the isentropic process

Minimizing Work Required by a Compressor

Comparing Reversible to Irreversible Compressors

Comparison of the differential form of the energy balance for each deviceLet q be heat, w be work, ke be kinetic energy and pe be potential energy.Actual Compressor:

$$\delta q_{act} - \delta w_{act} = dh + dke + dpe$$

Reversible Compressor:

$$\delta q_{rev} - \delta w_{rev} = dh + dke + dpe$$

The right hand side of each compressor type is equivalent, thus:

$$\delta q_{act} - \delta w_{act} = \delta q_{rev} - \delta w_{rev}$$

re-arranging:

$$\delta w_{rev} - \delta w_{act} = \delta q_{rev} - \delta q_{act}$$

By substituting the know equation $\delta q_{rev} = Tds$ into the last equation and dividing both terms by T:

$$\frac{\delta w_{rev} - \delta w_{act}}{T} = ds - \frac{\delta q_{act}}{T} \geq 0$$

Furthermore, $ds \geq \dfrac{\delta q_{act}}{T}$ and T is [absolute temperature] $(T \geq 0)$ which produces:

$$\delta w_{rev} \geq \delta w_{act}$$

or

$$w_{rev} \geq w_{act}$$

Therefore, work-consuming devices such as pumps and compressors (work is negative) require less work when they operate reversibly.

Effect of Cooling During the Compression Process

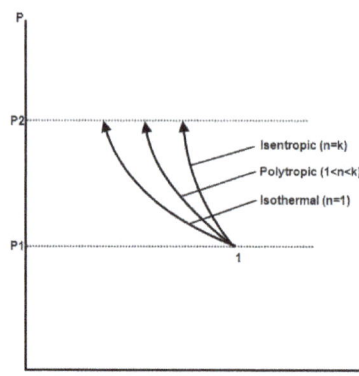

P-v (Specific volume vs. Pressure) diagram comparing isentropic, polytropic, and isothermal processes between the same pressure limits.

isentropic process: involves no cooling, polytropic process: involves some cooling isothermal process: involves maximum cooling

By making the following assumptions the required work for the compressor to compress a gas from P_1 to P_2 is the following for each process:Assumptions:

P_1 and P_2

All processes are internally reversible

The gas behaves like an ideal gas with constant specific heats

Isentropic ($Pv^k = constant$, where $k = C_p / C_v$):

$$W_{comp,in} = \frac{kR(T_2 - T_1)}{k-1} = \frac{kRT_1}{k-1}\left[\left(\frac{P_2}{P_1}\right)^{(k-1)/k} - 1\right]$$

Polytropic ($Pv^n = constant$):

$$W_{comp,in} = \frac{nR(T_2 - T_1)}{n-1} = \frac{nRT_1}{n-1}\left[\left(\frac{P_2}{P_1}\right)^{(n-1)/n} - 1\right]$$

Isothermal ($T = constant$ or $Pv = constant$):

$$W_{comp,in} = RTln\left(\frac{P_2}{P_1}\right)$$

By comparing the three internally reversible processes compressing an ideal gas from P_1 to P_2, the results show that isentropic compression ($Pv^k = constant$) requires the most work in and the isothermal compression($T = constant$ or $Pv = constant$) requires the least amount of work in. For the polytropic process ($Pv^n = constant$) work in decreases as the exponent, n, decreases, by increasing the heat rejection during the compression process. One common way of cooling the gas during compression is to use cooling jackets around the casing of the compressor.

Compressors in Ideal Thermodynamic Cycles

Ideal Rankine Cycle 1->2 Isentropic compression in a pump

Ideal Carnot Cycle 4->1 Isentropic compression

Ideal Otto Cycle 1->2 Isentropic compression

Ideal Diesel Cycle 1->2 Isentropic compression

Ideal Brayton Cycle 1->2 Isentropic compression in a compressor

Ideal Vapor-compression refrigeration Cycle 1->2 Isentropic compression in a compressor NOTE: The isentropic assumptions are only applicable with ideal cycles. Real world cycles have inherent losses due to inefficient compressors and turbines. The real world system are not truly isentropic but are rather idealized as isentropic for calculation purposes.

Temperature

Compression of a gas increases its temperature.

$$W = \int_{V_1}^{V_2} p\,dV = p_1 V_1^n \int_{V_1}^{V_2} V^{-n}\,dV$$

where

$$\frac{p_2}{p_1} = \left(\frac{V_1}{V_2}\right)^n$$

or

$$p_1 V_1^n = p_2 V_2^n = pV^n$$

and

$$p = \frac{p_1 V_1^n}{V^n}$$

so

$$W = \frac{p_1 V_1^n}{1-n}(V_2^{1-n} - V_1^{1-n})$$

in which p is pressure, V is volume, n takes different values for different compression processes, and 1 & 2 refer to initial and final states.

- Adiabatic - This model assumes that no energy (heat) is transferred to or from the gas during the compression, and all supplied work is added to the internal energy of the gas, resulting in increases of temperature and pressure. Theoretical temperature rise is:

$$T_2 = T_1 \left(\frac{p_2}{p_1}\right)^{(k-1)/k}$$

with T_1 and T_2 in degrees Rankine or kelvins, p_2 and p_1 being absolute pressures and k = ratio of specific heats (approximately 1.4 for air). The rise in air and temperature ratio means compression does not follow a simple pressure to volume ratio. This is less efficient, but quick. Adiabatic compression or expansion more closely model real life when a compressor has good insulation, a large gas volume, or a short time scale (i.e., a high power level). In practice there will always be a certain amount of heat flow out of the compressed gas. Thus, making a perfect adiabatic compressor would require perfect heat insulation of all parts of the machine. For example, even a bicycle tire pump's metal tube becomes hot as you compress the air to fill a tire. The relation between temperature and compression ratio described above means that the value of n for an adiabatic process is k (the ratio of specific heats).

- Isothermal - This model assumes that the compressed gas remains at a constant temperature throughout the compression or expansion process. In this cycle, internal energy is removed from the system as heat at the same rate that it is added by the mechanical work of compression. Isothermal compression or expansion more closely models real life when the compressor has a large heat exchanging surface, a small gas volume, or a long time scale (i.e., a small power level). Compressors that utilize inter-stage cooling between compression stages come closest to achieving perfect isothermal compression. However, with practical devices perfect isothermal compression is not attainable. For example, unless you have an infinite number of compression stages with corresponding intercoolers, you will never achieve perfect isothermal compression.

For an isothermal process, n is 1, so the value of the work integral for an isothermal process is:

$$W = -p_1 V_1 \ln\left(\frac{p_2}{p_1}\right)$$

When evaluated, the isothermal work is found to be lower than the adiabatic work.

- Polytropic - This model takes into account both a rise in temperature in the gas as well as some loss of energy (heat) to the compressor's components. This assumes that heat may enter or leave the system, and that input shaft work can appear as both increased pressure (usually useful work) and increased temperature above adiabatic (usually losses due to cycle efficiency). Compression efficiency is then the ratio of temperature rise at theoretical 100 percent (adiabatic) vs. actual (polytropic). Polytropic compression will use a value of n between 0 (a constant-pressure process) and infinity (a constant volume process). For the typical case where an effort is made to cool the gas compressed by an approximately adiabatic process, the value of n will be between 1 and k.

Staged Compression

In the case of centrifugal compressors, commercial designs currently do not exceed a compression ratio of more than a 3.5 to 1 in any one stage (for a typical gas). Since compression raises the temperature, the compressed gas is to be cooled between stages making the compression less adiabatic and more isothermal. The inter-stage coolers typically result in some partial condensation that is removed in vapor-liquid separators.

In the case of small reciprocating compressors, the compressor flywheel may drive a cooling fan that directs ambient air across the intercooler of a two or more stage compressor.

Because rotary screw compressors can make use of cooling lubricant to reduce the temperature rise from compression, they very often exceed a 9 to 1 compression ratio. For instance, in a typical diving compressor the air is compressed in three stages. If each stage has a compression ratio of 7 to 1, the compressor can output 343 times atmospheric pressure ($7 \times 7 \times 7 = 343$ atmospheres). (343 atm or 34.8 MPa or 5.04 ksi)

Drive Motors

There are many options for the motor that powers the compressor:

- Gas turbines power the axial and centrifugal flow compressors that are part of jet engines.

- Steam turbines or water turbines are possible for large compressors.

- Electric motors are cheap and quiet for static compressors. Small motors suitable for domestic electrical supplies use single-phase alternating current. Larger motors can only be used where an industrial electrical three phase alternating current supply is available.

- Diesel engines or petrol engines are suitable for portable compressors and support compressors.

- In automobiles and other types of vehicles (including piston-powered airplanes, boats, trucks, etc.), diesel or gasoline engines power output can be increased by compressing the intake air, so that more fuel can be burned per cycle. These engines can power compressors using their own crankshaft power (this setup known as a supercharger), or, use their exhaust gas to drive a turbine connected to the compressor (this setup known as a turbocharger).

Applications

Gas compressors are used in various applications where either higher pressures or lower volumes of gas are needed:

- In pipeline transport of purified natural gas from the production site to the consumer, a compressor is driven by a gas turbine fueled by gas bled from the pipeline. Thus, no external power source is necessary.

- Petroleum refineries, natural gas processing plants, petrochemical and chemical plants, and similar large industrial plants require compressing for intermediate and end-product gases.

- Refrigeration and air conditioner equipment use compressors to move heat in refrigerant cycles.

- Gas turbine systems compress the intake combustion air.

- Small-volume purified or manufactured gases require compression to fill high pressure cylinders for medical, welding, and other uses.

- Various industrial, manufacturing, and building processes require compressed air to power pneumatic tools.

- In the manufacturing and blow moulding of PET plastic bottles and containers.

- Some aircraft require compressors to maintain cabin pressurization at altitude.

- Some types of jet engines—such as turbojets and turbofans)—compress the air required for fuel combustion. The jet engine's turbines power the combustion air compressor.

- In SCUBA diving, hyperbaric oxygen therapy, and other life support devices, compressors put breathing gas into small volume containers, such as diving cylinders.

- In surface supplied diving, an air compressor frequently supplies low pressure air (10 to 20 bar) for breathing.

- Submarines use compressors to store air for later use in displacing water from buoyancy chambers to adjust depth.

- Turbochargers and superchargers are compressors that increase internal combustion engine performance by increasing the mass flow of air inside the cylinder, so the engine can burn more fuel and hence produce more power.

- Rail and heavy road transport vehicles use compressed air to operate rail vehicle or road vehicle brakes—and various other systems (doors, windscreen wipers, engine, gearbox control, etc.).

- Service stations and auto repair shops use compressed air to fill pneumatic tires and power pneumatic tools.

- Fire pistons and heat pumps exist to heat air or other gasses, and compressing the gas is only a means to that end.

Diving air compressor in noise reduction cabinet

Combustion Chamber

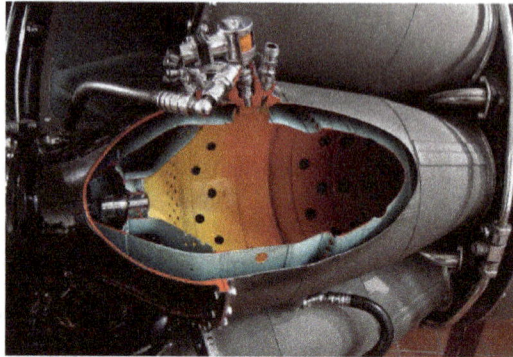

Combustion chamber (named combustor) on a Rolls-Royce Nene turbojet engine

A combustion chamber is that part of an internal combustion engine (ICE) in which the fuel/air mix is burned.

Internal Combustion Engine

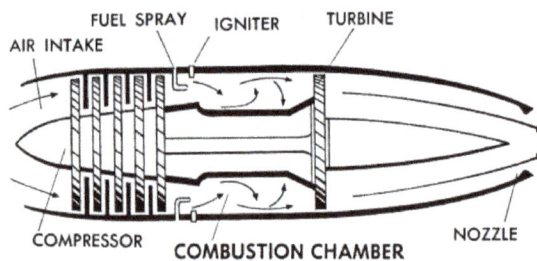

Diagram of jet engine showing the combustion chamber.

ICEs typically comprise reciprocating piston engines, rotary engines, gas turbine and jet turbines.

The combustion process increases the internal energy of a gas, which translates into an increase in temperature, pressure, or volume depending on the configuration. In an enclosure, for example the cylinder of a reciprocating engine, the volume is controlled and the combustion creates an increase in pressure. In a continuous flow system, for example a jet engine combustor, the pressure is controlled and the combustion creates an increase in volume. This increase in pressure or volume can be used to do work, for example, to move a piston on a crankshaft or a turbine disc in a gas turbine. If the gas velocity changes, thrust is produced, such as in the nozzle of a rocket engine.

Petrol (Gasoline) Engine

At top dead centre the pistons of a petrol engine are flush (or nearly flush) with the top of the cylinder block. The combustion chamber may be a recess either in the cylinder head, or in the top of the piston. A design with the combustion chamber in the piston is called a Heron head, where the head is machined flat but the pistons are dished. The Heron head has proved even more thermodynamically efficient than the hemispherical head. Intake valves permit the inflow of a fuel air mix; and exhaust valves allow burnt gases to be scavenged.

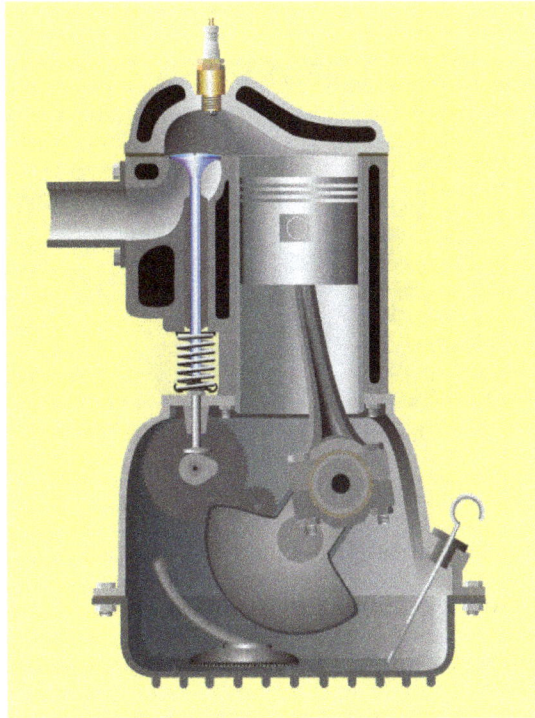

Side-valve engine showing combustion chamber

Head Types

Various shapes of combustion chamber have been used, such as: L-head (or flathead) for side-valve engines; "bathtub", "hemispherical", and "wedge" for overhead valve engines; and "pent-roof" for engines having 3, 4 or 5 valves per cylinder. The shape of the chamber has a marked effect on power output, efficiency and emissions; the designer's objectives are to burn all of the mixture as completely as possible while avoiding excessive temperatures (which create NOx). This is best achieved with a compact rather than elongated chamber.

Swirl & Squish

The intake valve/port is usually placed to give the mixture a pronounced "swirl" (the term is preferable to "turbulence", which implies movement without overall pattern) above the rising piston, improving mixing and combustion. The shape of the piston top also affects the amount of swirl. Another design feature to promote turbulence for good fuel/air mixing is "squish", where the fuel/air mix is "squished" at high pressure by the rising piston. Where swirl is particularly important, combustion chambers in the piston may be favoured.

Flame Front

Ignition typically occurs around 15 degrees before top dead centre. The spark plug must be sited so that the flame front can progress throughout the combustion chamber. Good design should avoid narrow crevices where stagnant "end gas" can become trapped, as this gas may detonate violently after the main charge, adding little useful work and potentially damaging the engine.

Diesel Engine

Dished piston for diesel engine

Diesel engines fall into two broad classes:

- Direct injection, where the combustion chamber consists of a dished piston
- Indirect injection, where the combustion chamber is in the cylinder head

Direct injection engines usually give better fuel economy but indirect injection engines can use a lower grade of fuel.

Harry Ricardo was prominent in developing combustion chambers for diesel engines, the best known being the Ricardo Comet.

Gas Turbine

The combustion chamber in gas turbines and jet engines (including ramjets and scramjets) is called the combustor.

The combustor is fed with high pressure air by the compression system, adds fuel and burns the mix and feeds the hot, high pressure exhaust into the turbine components of the engine or out the exhaust nozzle.

Different types of combustors exist, mainly:

- Can type: Can combustors are self-contained cylindrical combustion chambers. Each "can" has its own fuel injector, liner,interconnectors,casing. Each "can" get an air source from individual opening.

- Cannular type: Like the can type combustor, can annular combustors have discrete combustion zones contained in separate liners with their own fuel injectors. Unlike the can combustor, all the combustion zones share a common air casing.

- Annular type: Annular combustors do away with the separate combustion zones and simply have a continuous liner and casing in a ring (the annulus).

Rocket Engine

The term combustion chamber is also used to refer to an additional space between the firebox and boiler in a steam locomotive. This space is used to allow further combustion of the fuel, providing greater heat to the boiler.

Large steam locomotives usually have a combustion chamber in the boiler to allow the use of shorter firetubes. This is because:

Long firetubes have a theoretical advantage in providing a large heating surface but, beyond a certain length, this is subject to diminishing returns.

Very long firetubes are prone to sagging in the middle.

Micro Combustion Chambers

Micro combustion chambers are the devices in which combustion happens at a very small volume, due to which surface to volume ratio increases which plays a vital role in stabilizing the flame.

References

- Perry, R.H. and Green, D.W. (Editors) (2007). Perry's Chemical Engineers' Handbook (8th ed.). McGraw Hill. ISBN 0-07-142294-3.

- Bloch, H.P. and Hoefner, J.J. (1996). Reciprocating Compressors, Operation and Maintenance. Gulf Professional Publishing. ISBN 0-88415-525-0.

- Dixon S.L. (1978). Fluid Mechanics, Thermodynamics of Turbomachinery (Third ed.). Pergamon Press. ISBN 0-08-022722-8.

- Aungier, Ronald H. (2000). Centrifugal Compressors A Strategy for Aerodynamic design and Analysis. ASME Press. ISBN 0-7918-0093-8.

- Eric Slack (Winter 2016). "Sertco". Energy and Mining International. Phoenix Media Corporation. Retrieved February 27, 2016.

3

Types of Gas Turbines

Gas turbines are classified by the fuel they use and the thrust they produce. Different types of gas turbines seek to maximize the form of energy utilized. This chapter focuses on various types of gas turbines like turboprop, jet engine, auxiliary power unit, turboshaft, closed-cycle gas turbine, radial turbine, turbopump etc. providing valuable information about their technological aspects, uses, application and their types.

Jet Engine

A Pratt & Whitney F100 turbofan engine for the F-15 Eagle being tested in the hush house at Florida Air National Guard base. The tunnel behind the engine muffles noise and allows exhaust to escape

A jet engine is a reaction engine discharging a fast-moving jet that generates thrust by jet propulsion. This broad definition includes turbojets, turbofans, rocket engines, ramjets, and pulse jets. In general, jet engines are combustion engines.

In common parlance, the term *jet engine* loosely refers to an internal combustion airbreathing jet engine. These typically feature a rotating air compressor powered by a turbine, with the leftover power providing thrust via a propelling nozzle — this process is known as the Brayton thermodynamic cycle. Jet aircraft use such engines for long-distance travel. Early jet aircraft used turbojet engines which were relatively inefficient for subsonic flight. Modern subsonic jet aircraft usually use more complex high-bypass turbofan engines. These engines offer high speed and greater fuel efficiency than piston and propeller aeroengines over long distances.

U.S. Air Force F-15E Strike Eagles

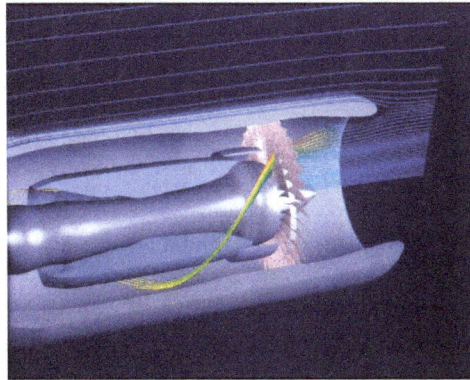

Simulation of a low-bypass turbofan's airflow.

Jet engine airflow during take-off.

The thrust of a typical jetliner engine went from 5,000 lbf (22,000 N) (de Havilland Ghost turbojet) in the 1950s to 115,000 lbf (510,000 N) (General Electric GE90 turbofan) in the 1990s, and their reliability went from 40 in-flight shutdowns per 100,000 engine flight hours to less than one in the late 1990s. This, combined with greatly decreased fuel consumption, permitted routine transatlantic flight by twin-engined airliners by the turn of the century, where before a similar journey would have required multiple fuel stops.

History

Jet engines date back to the invention of the aeolipile before the first century AD. This device directed steam power through two nozzles to cause a sphere to spin rapidly on its axis. So far as is known, it did not supply mechanical power and the potential practical applications of this invention did not receive recognition. Instead, it was seen as a curiosity.

Jet propulsion only gained practical applications with the invention of the gunpowder-powered rocket by the Chinese in the 13th century as a type of firework, and gradually progressed to propel formidable weaponry. However, although very powerful, at reasonable flight speeds rockets are very inefficient and so jet propulsion technology stalled for hundreds of years.

The earliest attempts at airbreathing jet engines were hybrid designs in which an external power source first compressed air, which was then mixed with fuel and burned for jet thrust. In one such system, called a *thermojet* by Secondo Campini but more commonly, motorjet, the air was compressed by a fan driven by a conventional piston engine. Examples of this type of design were the Caproni Campini N.1, and the Japanese Tsu-11 engine intended to power Ohka kamikaze planes towards the end of World War II. None were entirely successful and the N.1 ended up being slower than the same design with a traditional engine and propeller combination.

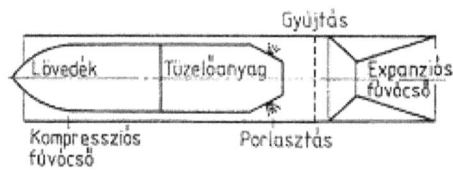

Albert Fonó's ramjet-cannonball from 1915

Even before the start of World War II, engineers were beginning to realize that engines driving propellers were self-limiting in terms of the maximum performance which could be attained; the limit was due to issues related to propeller efficiency, which declined as blade tips approached the speed of sound. If aircraft performance were ever to increase beyond such a barrier, a way would have to be found to use a different propulsion mechanism. This was the motivation behind the development of the gas turbine engine, commonly called a "jet" engine.

The key to a practical jet engine was the gas turbine, used to extract energy from the engine itself to drive the compressor. The gas turbine was not an idea developed in the 1930s: the patent for a stationary turbine was granted to John Barber in England in 1791. The first gas turbine to successfully run self-sustaining was built in 1903 by Norwegian engineer Ægidius Elling. Limitations in design and practical engineering and metallurgy prevented such engines reaching manufacture. The main problems were safety, reliability, weight and, especially, sustained operation.

The first patent for using a gas turbine to power an aircraft was filed in 1921 by Frenchman Maxime Guillaume. His engine was an axial-flow turbojet. Alan Arnold Griffith published *An Aerodynamic Theory of Turbine Design* in 1926 leading to experimental work at the RAE.

The Whittle W.2/700 engine flew in the Gloster E.28/39, the first British aircraft to fly with a turbojet engine, and the Gloster Meteor

In 1928, RAF College Cranwell cadet Frank Whittle formally submitted his ideas for a turbojet to his superiors. In October 1929 he developed his ideas further. On 16 January 1930 in England, Whittle submitted his first patent (granted in 1932). The patent showed a two-stage axial compressor feeding a single-sided centrifugal compressor. Practical axial compressors were made possible by ideas from A.A.Griffith in a seminal paper in 1926 ("An Aerodynamic Theory of Turbine Design"). Whittle would later concentrate on the simpler centrifugal compressor only, for a variety of practical reasons. Whittle had his first engine running in April 1937. It was liquid-fuelled, and included a self-contained fuel pump. Whittle's team experienced near-panic when the engine would not stop, accelerating even after the fuel was switched off. It turned out that fuel had leaked into the engine and accumulated in pools, so the engine would not stop until all the leaked fuel had burned off. Whittle was unable to interest the government in his invention, and development continued at a slow pace.

Heinkel He 178, the world's first aircraft to fly purely on turbojet power

In 1935 Hans von Ohain started work on a similar design in Germany, initially unaware of Whittle's work.

Von Ohain's first device was strictly experimental and could run only under external power, but he was able to demonstrate the basic concept. Ohain was then introduced to Ernst Heinkel, one of the larger aircraft industrialists of the day, who immediately saw the promise of the design. Heinkel had recently purchased the Hirth engine company, and Ohain and his master machinist Max Hahn were set up there as a new division of the Hirth company. They had their first HeS 1 centrifugal engine running by September 1937. Unlike Whittle's design, Ohain used hydrogen as fuel, supplied under external pressure. Their subsequent designs culminated in the gasoline-fuelled HeS 3 of 5 kN (1,100 lbf), which was fitted to Heinkel's simple and compact He 178 airframe and flown by Erich Warsitz in the early morning of August 27, 1939, from Rostock-Marienehe aerodrome, an impressively short time for development. The He 178 was the world's first jet plane.

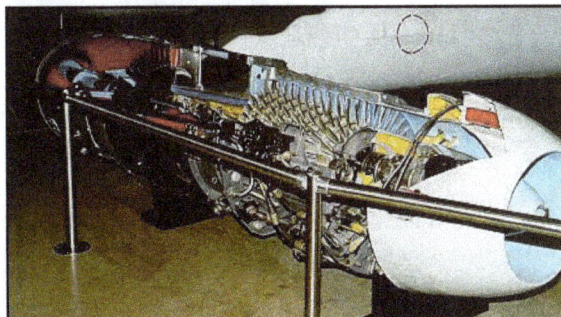

A cutaway of the Junkers Jumo 004 engine

Austrian Anselm Franz of Junkers' engine division (*Junkers Motoren* or "Jumo") introduced the axial-flow compressor in their jet engine. Jumo was assigned the next engine number in the RLM 109-0xx numbering sequence for gas turbine aircraft powerplants, "004", and the result was the Jumo 004 engine. After many lesser technical difficulties were solved, mass production of this engine started in 1944 as a powerplant for the world's first jet-fighter aircraft, the Messerschmitt Me 262 (and later the world's first jet-bomber aircraft, the Arado Ar 234). A variety of reasons conspired to delay the engine's availability, causing the fighter to arrive too late to improve Germany's position in World War II. Nonetheless, it will be remembered as the first use of jet engines in service.

Meanwhile, in Britain the Gloster E28/39 had its maiden flight on 15 May 1941 and the Gloster Meteor finally entered service with the RAF in July 1944. These were powered by turbojet engines from Power Jets Ltd., set up by Frank Whittle.

Following the end of the war the German jet aircraft and jet engines were extensively studied by the victorious allies and contributed to work on early Soviet and US jet fighters. The legacy of the axial-flow engine is seen in the fact that practically all jet engines on fixed-wing aircraft have had some inspiration from this design.

By the 1950s the jet engine was almost universal in combat aircraft, with the exception of cargo, liaison and other specialty types. By this point some of the British designs were already cleared for civilian use, and had appeared on early models like the de Havilland Comet and Avro Canada Jetliner. By the 1960s all large civilian aircraft were also jet powered, leaving the piston engine in low-cost niche roles such as cargo flights.

The efficiency of turbojet engines was still rather worse than piston engines, but by the 1970s, with the advent of high-bypass turbofan jet engines (an innovation not foreseen by the early commentators such as Edgar Buckingham, at high speeds and high altitudes that seemed absurd to them), fuel efficiency was about the same as the best piston and propeller engines.

Uses

A JT9D turbofan jet engine installed on a Boeing 747 aircraft.

Jet engines power jet aircraft, cruise missiles and unmanned aerial vehicles. In the form of rocket engines they power fireworks, model rocketry, spaceflight, and military missiles.

Jet engines have propelled high speed cars, particularly drag racers, with the all-time record held by a rocket car. A turbofan powered car, ThrustSSC, currently holds the land speed record.

Jet engine designs are frequently modified for non-aircraft applications, as industrial gas turbines or marine powerplants. These are used in electrical power generation, for powering water, natural gas, or oil pumps, and providing propulsion for ships and locomotives. Industrial gas turbines can create up to 50,000 shaft horsepower. Many of these engines are derived from older military turbojets such as the Pratt & Whitney J57 and J75 models. There is also a derivative of the P&W JT8D low-bypass turbofan that creates up to 35,000 HP.

Jet engines are also sometimes developed into, or share certain components such as engine cores, with turboshaft and turboprop engines, which are forms of gas turbine engines that are typically used to power helicopters and some propeller-driven aircraft..

Types

There are a large number of different types of jet engines, all of which achieve forward thrust from the principle of *jet propulsion.*

Airbreathing

Commonly aircraft are propelled by airbreathing jet engines. Most airbreathing jet engines that are in use are turbofan jet engines, which give good efficiency at speeds just below the speed of sound.

Turbine Powered

Gas turbines are rotary engines that extract energy from a flow of combustion gas. They have an upstream compressor coupled to a downstream turbine with a combustion chamber in-between. In aircraft engines, those three core components are often called the "gas generator." There are many different variations of gas turbines, but they all use a gas generator system of some type.

Turbojet

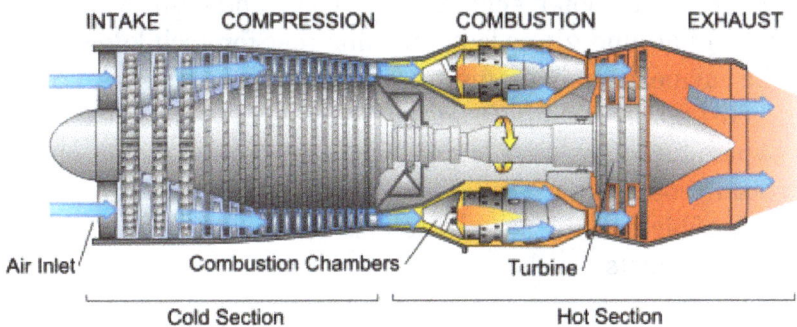

Turbojet engine

A turbojet engine is a gas turbine engine that works by compressing air with an inlet and a compressor (axial, centrifugal, or both), mixing fuel with the compressed air, burning the mixture in the combustor, and then passing the hot, high pressure air through a turbine and a nozzle.

The compressor is powered by the turbine, which extracts energy from the expanding gas passing through it. The engine converts internal energy in the fuel to kinetic energy in the exhaust, producing thrust. All the air ingested by the inlet is passed through the compressor, combustor, and turbine, unlike the turbofan engine described below.

Turbofan

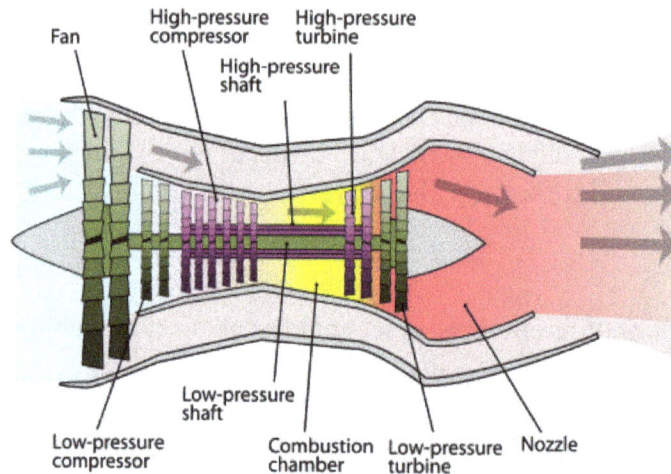

Schematic diagram illustrating the operation of a low-bypass turbofan engine.

A turbofan engine is a gas turbine engine that is very similar to a turbojet. Like a turbojet, it uses the gas generator core (compressor, combustor, turbine) to convert internal energy in fuel to kinetic energy in the exhaust. Turbofans differ from turbojets in that they have an additional component, a fan. Like the compressor, the fan is powered by the turbine section of the engine. Unlike the turbojet, some of the flow accelerated by the fan bypasses the gas generator core of the engine and is exhausted through a nozzle. The bypassed flow is at lower velocities, but a higher mass, making thrust produced by the fan more efficient than thrust produced by the core. Turbofans are generally more efficient than turbojets at subsonic speeds, but they have a larger frontal area which generates more drag.

There are two general types of turbofan engines, low-bypass and high-bypass. Low-bypass turbofans have a bypass ratio of around 2:1 or less, meaning that for each kilogram of air that passes through the core of the engine, two kilograms or less of air bypass the core. Low-bypass turbofans often use a mixed exhaust nozzle meaning that the bypassed flow and the core flow exit from the same nozzle. High-bypass turbofans have larger bypass ratios, sometimes on the order of 5:1 or 6:1. These turbofans can produce much more thrust than low-bypass turbofans or turbojets because of the large mass of air that the fan can accelerate, and are often more fuel efficient than low-bypass turbofans or turbojets.

Turboprop and Turboshaft

Turboprop engines are jet engine derivatives, still gas turbines, that extract work from the hot-exhaust jet to turn a rotating shaft, which is then used to produce thrust by some other means. While not strictly jet engines in that they rely on an auxiliary mechanism to produce thrust, turboprops are very similar to other turbine-based jet engines, and are often described as such.

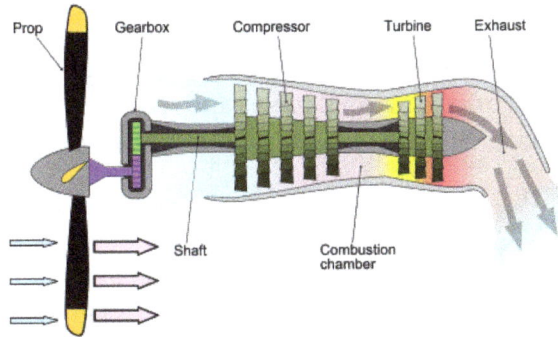

Turboprop engine

In turboprop engines, a portion of the engine's thrust is produced by spinning a propeller, rather than relying solely on high-speed jet exhaust. As their jet thrust is augmented by a propeller, turboprops are occasionally referred to as a type of hybrid jet engine. They are quite similar to turbofans in many respects, except that they use a traditional propeller to provide the majority of thrust, rather than a ducted fan. Both fans and propellers are powered the same way, although most turboprops use gear-reduction between the turbine and the propeller (geared turbofans also feature gear reduction). While many turboprops generate the majority of their thrust with the propeller, the hot-jet exhaust is an important design point, and maximum thrust is obtained by matching thrust contributions of the propeller to the hot jet. Turboprops generally have better performance than turbojets or turbofans at low speeds where propeller efficiency is high, but become increasingly noisy and inefficient at high speeds.

Turboshaft engines are very similar to turboprops, differing in that nearly all energy in the exhaust is extracted to spin the rotating shaft, which is used to power machinery rather than a propeller, they therefore generate little to no jet thrust and are often used to power helicopters.

Propfan

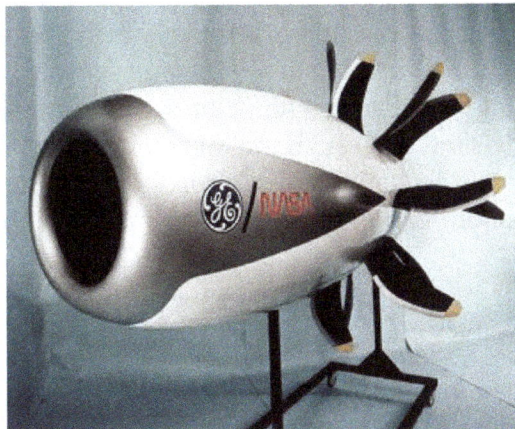

A propfan engine

A propfan engine (also called "unducted fan", "open rotor", or "ultra-high bypass") is a jet engine that uses its gas generator to power an exposed fan, similar to turboprop engines. Like turboprop engines, propfans generate most of their thrust from the propeller and not the exhaust jet. The primary difference between turboprop and propfan design is that the propeller blades on a propfan are highly swept to allow them to operate at speeds around Mach 0.8, which is competitive with modern commercial turbofans. These engines have the fuel efficiency advantages of turboprops with the performance capability of commercial turbofans. While significant research and testing (including flight testing) has been conducted on propfans, no propfan engines have entered production.

Ram Powered

Ram powered jet engines are airbreathing engines similar to gas turbine engines and they both follow the Brayton cycle. Gas turbine and ram powered engines differ, however, in how they compress the incoming airflow. Whereas gas turbine engines use axial or centrifugal compressors to compress incoming air, ram engines rely only on air compressed through the inlet or diffuser. Ram powered engines are considered the most simple type of air breathing jet engine because they can contain no moving parts.

Ramjet

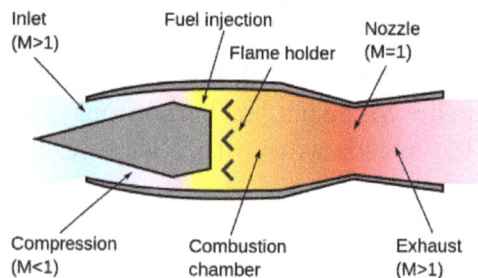

A schematic of a ramjet engine, where "M" is the Mach number of the airflow.

Ramjets are the most basic type of ram powered jet engines. They consist of three sections; an inlet to compress incoming air, a combustor to inject and combust fuel, and a nozzle to expel the hot gases and produce thrust. Ramjets require a relatively high speed to efficiently compress the incoming air, so ramjets cannot operate at a standstill and they are most efficient at supersonic speeds. A key trait of ramjet engines is that combustion is done at subsonic speeds. The supersonic incoming air is dramatically slowed through the inlet, where it is then combusted at the much slower, subsonic, speeds. The faster the incoming air is, however, the less efficient it becomes to slow it to subsonic speeds. Therefore, ramjet engines are limited to approximately Mach 5.

Scramjet

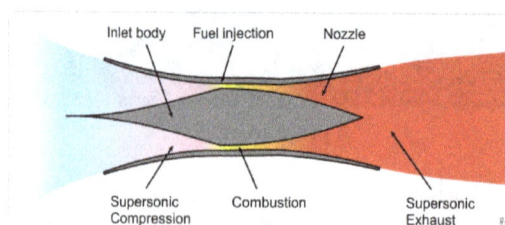

Scramjet engine operation

Scramjets are mechanically very similar to ramjets. Like a ramjet, they consist of an inlet, a combustor, and a nozzle. The primary difference between ramjets and scramjets is that scramjets do not slow the oncoming airflow to subsonic speeds for combustion, they use supersonic combustion instead. The name "scramjet" comes from "Supersonic Combusting Ramjet." Since scramjets use supersonic combustion they can operate at speeds above Mach 6 where traditional ramjets are too inefficient. Another difference between ramjets and scramjets comes from how each type of engine compresses the oncoming airflow: while the inlet provides most of the compression for ramjets, the high speeds at which scramjets operate allow them to take advantage of the compression generated by shock waves, primarily oblique shocks.

Very few scramjet engines have ever been built and flown. In May 2010 the Boeing X-51 set the endurance record for the longest scramjet burn at over 200 seconds.

Non-Continuous Combustion

Type	Description	Advantages	Disadvantages
Motorjet	Obsolete type that worked like a turbojet but instead of a turbine driving the compressor a piston engine drives it.	Higher exhaust velocity than a propeller, offering better thrust at high speed	Heavy, inefficient and underpowered. Example: Caproni Campini N.1.
Pulsejet	Air is compressed and combusted intermittently instead of continuously. Some designs use valves.	Very simple design, commonly used on model aircraft	Noisy, inefficient (low compression ratio), works poorly on a large scale, valves on valved designs wear out quickly
Pulse detonation engine	Similar to a pulsejet, but combustion occurs as a detonation instead of a deflagration, may or may not need valves	Maximum theoretical engine efficiency	Extremely noisy, parts subject to extreme mechanical fatigue, hard to start detonation, not practical for current use

Rocket

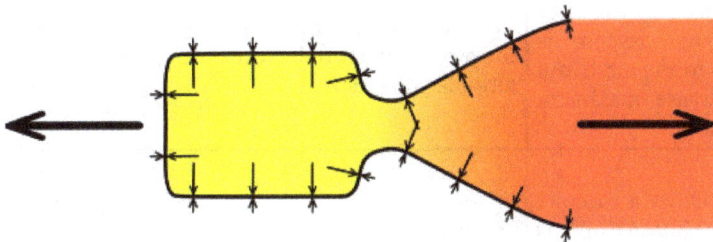

Rocket engine propulsion

The rocket engine uses the same basic physical principles as the jet engine for propulsion via thrust, but is distinct in that it does not require atmospheric air to provide oxygen; the rocket carries all components of the reaction mass. This allows them to operate at arbitrary altitudes and in space.

This type of engine is used for launching satellites, space exploration and manned access, and permitted landing on the moon in 1969.

Rocket engines are used for high altitude flights, or anywhere where very high accelerations are needed since rocket engines themselves have a very high thrust-to-weight ratio.

However, the high exhaust speed and the heavier, oxidizer-rich propellant results in far more propellant use than turbofans. Even so, at extremely high speeds they become energy-efficient.

An approximate equation for the net thrust of a rocket engine is:

$$F_N = \dot{m} g_0 I_{sp-vac} - A_e p$$

Where F_N is the net thrust, $I_{sp(vac)}$ is the specific impulse, g_0 is a standard gravity, $\dot{m}$ is the propellant flow in kg/s, A_e is the cross-sectional area at the exit of the exhaust nozzle, and p is the atmospheric pressure.

Type	Description	Advantages	Disadvantages
Rocket	Carries all propellants and oxidants on board, emits jet for propulsion	Very few moving parts. Mach 0 to Mach 25+; efficient at very high speed (> Mach 5.0 or so). Thrust/weight ratio over 100. No complex air inlet. High compression ratio. Very high-speed (hypersonic) exhaust. Good cost/thrust ratio. Fairly easy to test. Works in a vacuum; indeed, works best outside the atmosphere, which is kinder on vehicle structure at high speed. Fairly small surface area to keep cool, and no turbine in hot exhaust stream. Very high-temperature combustion and high expansion-ratio nozzle gives very high efficiency, at very high speeds.	Needs lots of propellant. Very low specific impulse—typically 100–450 seconds. Extreme thermal stresses of combustion chamber can make reuse harder. Typically requires carrying oxidizer on-board which increases risks. Extraordinarily noisy.

Hybrid

Combined cycle engines simultaneously use 2 or more different jet engine operating principles.

Type	Description	Advantages	Disadvantages
Turborocket	A turbojet where an additional oxidizer such as oxygen is added to the airstream to increase maximum altitude	Very close to existing designs, operates in very high altitude, wide range of altitude and airspeed	Airspeed limited to same range as turbojet engine, carrying oxidizer like LOX can be dangerous. Much heavier than simple rockets.
Air-augmented rocket	Essentially a ramjet where intake air is compressed and burnt with the exhaust from a rocket	Mach 0 to Mach 4.5+ (can also run exoatmospheric), good efficiency at Mach 2 to 4	Similar efficiency to rockets at low speed or exoatmospheric, inlet difficulties, a relatively undeveloped and unexplored type, cooling difficulties, very noisy, thrust/weight ratio is similar to ramjets.
Precooled jets / LACE	Intake air is chilled to very low temperatures at inlet in a heat exchanger before passing through a ramjet and/or turbojet and/or rocket engine.	Easily tested on ground. Very high thrust/weight ratios are possible (~14) together with good fuel efficiency over a wide range of airspeeds, Mach 0-5.5+; this combination of efficiencies may permit launching to orbit, single stage, or very rapid, very long distance intercontinental travel.	Exists only at the lab prototype stage. Examples include RB545, Reaction Engines SABRE, ATREX. Requires liquid hydrogen fuel which has very low density and requires heavily insulated tankage.

Water Jet

A water jet, or pump jet, is a marine propulsion system that utilizes a jet of water. The mechanical arrangement may be a ducted propeller with nozzle, or a centrifugal compressor and nozzle.

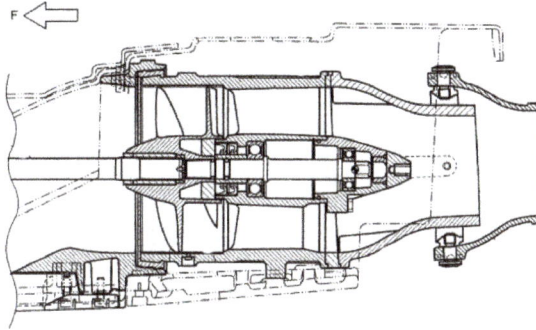

A pump jet schematic.

Type	Description	Advantages	Disadvantages
Water jet	For propelling water rockets and jetboats; squirts water out the back through a nozzle	In boats, can run in shallow water, high acceleration, no risk of engine overload (unlike propellers), less noise and vibration, highly maneuverable at all boat speeds, high speed efficiency, less vulnerable to damage from debris, very reliable, more load flexibility, less harmful to wildlife	Can be less efficient than a propeller at low speed, more expensive, higher weight in boat due to entrained water, will not perform well if boat is heavier than the jet is sized for

General Physical Principles

All jet engines are reaction engines that generate thrust by emitting a jet of fluid rearwards at relatively high speed. The forces on the inside of the engine needed to create this jet give a strong thrust on the engine which pushes the craft forwards.

Jet engines make their jet from propellant from tankage that is attached to the engine (as in a 'rocket') as well as in duct engines (those commonly used on aircraft) by ingesting an external fluid (very typically air) and expelling it at higher speed.

Propelling Nozzle

The propelling nozzle is the key component of all jet engines as it creates the exhaust jet. Propelling nozzles turn internal and pressure energy into high velocity kinetic energy. The total pressure and temperature don't change through the nozzle but their static values drop as the gas speeds up.

The velocity of the air entering the nozzle is low, about Mach 0.4, a prerequisite for minimising pressure losses in the duct leading to the nozzle. The temperature entering the nozzle may be as low as sea level ambient for a fan nozzle in the cold air at cruise altitudes. It may be as high as the 1000K exhaust gas temperature for a supersonic afterburning engine or 2200K with afterburner lit. The pressure entering the nozzle may vary from 1.5 times the pressure outside the nozzle, for a single stage fan, to 30 times for the fastest manned aircraft at mach 3+.

The velocity of the gas leaving a convergent nozzle may be subsonic or sonic (Mach 1) at low flight speeds or supersonic (Mach 3.0 at SR-71 cruise) for a con-di nozzle at higher speeds where the nozzle pressure ratio is increased with the intake ram. The nozzle thrust is highest if the static pressure of the gas reaches the ambient value as it leaves the nozzle. This only happens if the nozzle exit area is the correct value for the nozzle pressure ratio (npr). Since the npr changes with engine thrust setting and flight speed this is seldom the case. Also at supersonic speeds the divergent area is less than required to give complete internal expansion to ambient pressure as a trade-off with external body drag. Whitford gives the F-16 as an example. Other underexpanded examples were the XB-70 and SR-71.

The nozzle size, together with the area of the turbine nozzles, determines the operating pressure of the compressor.

Thrust

Origin of Engine Thrust

The familiar explanation for jet thrust is a "black box" description which only looks at what goes in to the engine, air and fuel, and what comes out, exhaust gas and an unbalanced force. This force, called thrust, is the sum of the momentum difference between entry and exit and any unbalanced pressure force between entry and exit, as explained in "Thrust calculation". As an example, an early turbojet, the Bristol Olympus Mk. 101, had a momentum thrust of 9300 lb. and a pressure thrust of 1800 lb. giving a total of 11,100 lb. Looking inside the "black box" shows that the thrust results from all the unbalanced momentum and pressure forces created within the engine itself. These forces, some forwards and some rearwards, are across all the internal parts, both stationary and rotating, such as ducts, compressors, etc., which are in the primary gas flow which flows through the engine from front to rear. The algebraic sum of all these forces is delivered to the airframe for propulsion. "Flight" gives examples of these internal forces for two early jet engines, the Rolls-Royce Avon Ra.14 and the de Havilland Goblin

Transferring Thrust to The Aircraft

The engine thrust acts along the engine centreline. The aircraft "holds" the engine on the outer casing of the engine at some distance from the engine centreline (at the engine mounts). This arrangement causes the engine casing to bend (known as backbone bending) and the round rotor casings to distort (ovalization). Distortion of the engine structure has to be controlled with suitable mount locations to maintain acceptable rotor and seal clearances and prevent rubbing. A well-publicized example of excessive structural deformation occurred with the original Pratt & Whitney JT9D engine installation in the Boeing 747 aircraft. The engine mounting arrangement had to be revised with the addition of an extra thrust frame to reduce the casing deflections to an acceptable amount.

Rotor Thrust

The rotor thrust on a thrust bearing is not related to the engine thrust. It may even change direction at some RPM. The bearing load is determined by bearing life considerations. Although the aerodynamic loads on the compressor and turbine blades contribute to the rotor thrust they are small compared to cavity loads inside the rotor which result from the secondary air system pressures and sealing diameters on discs, etc. To keep the load within the bearing specification seal diameters are chosen accordingly as, many years ago, on the backface of the impeller in the de Havilland Ghost engine. Sometimes an extra disc known as a balance piston has to be added inside the rotor. An early turbojet example with a balance piston was the Rolls-Royce Avon.

Thrust Calculation

The net thrust (F_N) of a turbojet is given by:

$$F_N = (\dot{m}_{air} + \dot{m}_{fuel})v_e - \dot{m}_{air}v$$

where:	
$\dot{m}_{air}$	= the mass rate of air flow through the engine
$\dot{m}_{fuel}$	= the mass rate of fuel flow entering the engine
v_e	= the velocity of the jet (the exhaust plume) and is assumed to be less than sonic velocity
v	= the velocity of the air intake = the true airspeed of the aircraft
$(\dot{m}_{air} + \dot{m}_{fuel})v_e$	= the nozzle gross thrust (F_G)
$\dot{m}_{air}v$	= the ram drag of the intake air

The above equation applies only for air-breathing jet engines. It does not apply to rocket engines. Most types of jet engine have an air intake, which provides the bulk of the fluid exiting the exhaust. Conventional rocket engines, however, do not have an intake, the oxidizer and fuel both being carried within the vehicle. Therefore, rocket engines do not have ram drag and the gross thrust of the rocket engine nozzle is the net thrust of the engine. Consequently, the thrust characteristics of a rocket motor are different from that of an air breathing jet engine, and thrust is independent of velocity.

If the velocity of the jet from a jet engine is equal to sonic velocity, the jet engine's nozzle is said to be choked. If the nozzle is choked, the pressure at the nozzle exit plane is greater than atmospheric pressure, and extra terms must be added to the above equation to account for the pressure thrust.

The rate of flow of fuel entering the engine is very small compared with the rate of flow of air. If the contribution of fuel to the nozzle gross thrust is ignored, the net thrust is:

$$F_N = \dot{m}_{air}(v_e - v)$$

The velocity of the jet (v_e) must exceed the true airspeed of the aircraft (v) if there is to be a net forward thrust on the aircraft. The velocity (v_e) can be calculated thermodynamically based on adiabatic expansion.

Thrust Augmentation

Thrust augmentation has taken many forms, most commonly to supplement inadequate take-off thrust. Some early jet aircraft needed rocket assistance to take off from high altitude airfields or when the day temperature was high. A more recent aircraft, the Tupolev Tu-22 supersonic bomber, was fitted with four SPRD-63 boosters for take-off. Possibly the most extreme requirement needing rocket assistance, and which was short-lived, was zero-length launching. Almost as extreme, but very common, is catapult assistance from aircraft carriers. Rocket assistance has also been used during flight. The SEPR 841 booster engine was used on the Dassault Mirage for high altitude interception.

Early aft-fan arrangements which added bypass airflow to a turbojet were known as thrust augmentors. The aft-fan fitted to the General Electric CJ805-3 turbojet augmented the take-off thrust from 11,650lb to 16,100lb.

Water, or other coolant, injection into the compressor or combustion chamber and fuel injection into the jetpipe (afterburning/reheat) became standard ways to increase thrust, known as 'wet' thrust to differentiate with the no-augmentation 'dry' thrust.

Coolant injection (pre-compressor cooling) has been used, together with afterburning, to increase thrust at supersonic speeds. The 'Skyburner' McDonnell Douglas F-4 Phantom II set a world speed record using water injection in front of the engine.

At high Mach numbers afterburners supply progressively more of the engine thrust as the thrust from the turbomachine drops off towards zero at which speed the engine pressure ratio (epr) has fallen to 1.0 and all the engine thrust comes from the afterburner. The afterburner also has to make up for the pressure loss across the turbomachine which is a drag item at higher speeds where the epr will be less than 1.0.

Thrust augmentation of existing afterburning engine installations for special short-duration tasks has been the subject of studies for launching small payloads into low earth orbits using aircraft such as McDonnell Douglas F-4 Phantom II, McDonnell Douglas F-15 Eagle, Dassault Rafale and Mikoyan MiG-31, and also for carrying experimental packages to high altitudes using a Lockheed SR-71. In the first case an increase in the existing maximum speed capability is required for orbital launches. In the second case an increase in thrust within the existing speed capability is required. Compressor inlet cooling is used in the first case. A compressor map shows that the airflow reduces with increasing compressor inlet temperature although the compressor is still running at maximum RPM (but reduced aerodynamic speed). Compressor inlet cooling increases the aerodynamic speed and flow and thrust. In the second case a small increase in the maximum mechanical speed and turbine temperature were allowed, together with nitrous oxide injection into the afterburner and simultaneous increase in afterburner fuel flow.

Energy Efficiency Relating to Aircraft Jet Engines

This overview highlights where energy losses occur in complete jet aircraft powerplants or engine installations. It includes mention of inlet and exhaust nozzle losses which become increasingly significant at the high flight speeds achieved by some manned aircraft since only a small proportion, 17% for the SR-71 powerplant and 8% for the Concorde powerplant, of the thrust transmitted to the airframe came from the engine.

A jet engine at rest, as on a test stand, sucks in fuel and tries to thrust itself forward. How well it does this is judged by how much fuel it uses and what force is required to restrain it. This is a measure of its efficiency. If something deteriorates inside the engine (known as performance deterioration) it will be less efficient and this will show when the fuel produces less thrust. If a change is made to an internal part which allows the air/combustion gases to flow more smoothly the engine will be more efficient and use less fuel. A standard definition is used to assess how different things change engine efficiency and also to allow comparisons to be made between different engines. This definition is called specific fuel consumption, or how much fuel is needed to produce one unit of thrust. For example, it will be known for a particular engine design that if some bumps in a bypass duct are smoothed out the air will flow more smoothly giving a pressure loss reduction of x% and y% less fuel will be needed to get the take-off thrust, for example. This understanding comes under the engineering discipline Jet engine performance. How efficiency is affected by forward speed and by supplying energy to aircraft systems is mentioned later.

The efficiency of the engine is controlled primarily by the operating conditions inside the engine which are the pressure produced by the compressor and the temperature of the combustion gases at the first set of rotating turbine blades. The pressure is the highest air pressure in the engine. The turbine rotor temperature is not the highest in the engine but is the highest at which energy transfer takes place (higher temperatures occur in the combustor). The above pressure and temperature are shown on a Thermodynamic cycle diagram.

The efficiency is further modified by how smoothly the air and the combustion gases flow through the engine, how well the flow is aligned (known as incidence angle) with the moving and stationary passages in the compressors and turbines. Non-optimum angles, as well as non-optimum passage and blade shapes can cause thickening and separation of Boundary layers and formation of Shock waves as explained in Effects of Mach number and shock losses in turbomachines. It is important to slow the flow (lower speed means less pressure losses or Pressure drop) when it travels through ducts connecting the different parts. How well the individual components contribute to turning fuel into thrust is quantified by measures like efficiencies for the compressors, turbines and combustor and pressure losses for the ducts. These are shown as lines on a Thermodynamic cycle diagram.

The engine efficiency, or thermal efficiency, known as η_{th}. is dependent on the Thermodynamic cycle parameters, maximum pressure and temperature, and on component efficiencies, $\eta_{compressor}$, $\eta_{combustion}$ and $\eta_{turbine}$ and duct pressure losses.

The engine needs compressed air for itself just to run successfully. This air comes from its own compressor and is called secondary air. It does not contribute to making thrust so makes the engine less efficient. It is used to preserve the mechanical integrity of the engine, to stop parts over-heating and to prevent oil escaping from bearings for example. Only some of this air taken from the compressors returns to the turbine flow to contribute to thrust production. Any reduction in the amount needed improves the engine efficiency. Again, it will be known for a particular engine design that a reduced requirement for cooling flow of x% will reduce the specific fuel consumption by y%. In other words, less fuel will be required to give take-off thrust, for example. The engine is more efficient.

All of the above considerations are basic to the engine running on its own and, at the same time, doing nothing useful, i.e. it is not moving an aircraft or supplying energy for the aircraft's electrical, hydraulic and air systems. In the aircraft the engine gives away some of its thrust-producing potential, or fuel, to power these systems. These requirements, which cause installation losses, reduce its efficiency. It is using some fuel that does not contribute to the engine's thrust.

Finally, when the aircraft is flying the propelling jet itself contains wasted kinetic energy after it has left the engine. This is quantified by the term propulsive, or Froude, efficiency η_p and may be reduced by redesigning the engine to give it bypass flow and a lower speed for the propelling jet, for example as a turboprop or turbofan engine. At the same time forward speed increases the by increasing the Overall pressure ratio.

The overall efficiency of the engine at flight speed is defined as $\eta_o = \eta_p \eta_{th}$.

The η_o at flight speed depends on how well the intake compresses the air before it is handed over to the engine compressors. The intake compression ratio, which can be as high as 32:1 at Mach 3, adds to that of the engine compressor to give the Overall pressure ratio and η_{th} for the Thermo-dynamic cycle. How well it does this is defined by its pressure recovery or measure of the losses in the intake. Mach 3 manned flight has provided an interesting illustration of how these losses can increase dramatically in an instant. The North American XB-70 Valkyrie and Lockheed SR-71 Blackbird at Mach 3 each had pressure recoveries of about 0.8, due to relatively low losses during the compression process, i.e. through systems of multiple shocks. During an 'unstart' the efficient shock system would be replaced by a very inefficient single shock beyond the inlet and an intake pressure recovery of about 0.3 and a correspondingly low pressure ratio.

The propelling nozzle at speeds above about Mach 2 usually has extra internal thrust losses be-cause the exit area is not big enough as a trade-off with external afterbody drag.

Although a bypass engine improves propulsive efficiency it incurs losses of its own inside the en-gine itself. Machinery has to be added to transfer energy from the gas generator to a bypass airflow. The low loss from the propelling nozzle of a turbojet is added to with extra losses due to inefficien-cies in the added turbine and fan. These may be included in a transmission, or transfer, efficiency η_T. However, these losses are more than made up by the improvement in propulsive efficiency. There are also extra pressure losses in the bypass duct and an extra propelling nozzle.

With the advent of turbofans with their loss-making machinery what goes on inside the engine has been separated by Bennett, for example, between gas generator and transfer machinery giving $\eta_o = \eta_p \eta_{th} \eta_T$.

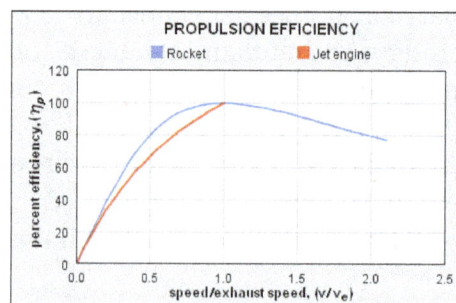

Dependence of propulsion efficiency (η) upon the vehicle speed/exhaust velocity ratio (v/v_e) for air-breathing jet and rocket engines.

The energy efficiency (η_o) of jet engines installed in vehicles has two main components:

- *propulsive efficiency* (η_p): how much of the energy of the jet ends up in the vehicle body rather than being carried away as kinetic energy of the jet.

- *cycle efficiency* ($_{th}$): how efficiently the engine can accelerate the jet

Even though overall energy efficiency η_o is:

$$\eta_o = \eta_p \eta_{th}$$

for all jet engines the *propulsive efficiency* is highest as the exhaust jet velocity gets closer to the vehicle speed as this gives the smallest residual kinetic energy. For an airbreathing engine an exhaust velocity equal to the vehicle velocity, or a η_p equal to one, gives zero thrust with no net momentum change. The formula for air-breathing engines moving at speed v with an exhaust velocity v_e, and neglecting fuel flow, is:

$$\eta_p = \frac{2}{1 + \dfrac{v_e}{v}}$$

And for a rocket:

$$\eta_p = \frac{2(\dfrac{v}{v_e})}{1 + (\dfrac{v}{v_e})^2}$$

In addition to propulsive efficiency, another factor is *cycle efficiency*; a jet engine is a form of heat engine. Heat engine efficiency is determined by the ratio of temperatures reached in the engine to that exhausted at the nozzle. This has improved constantly over time as new materials have been introduced to allow higher maximum cycle temperatures. For example, composite materials, combining metals with ceramics, have been developed for HP turbine blades, which run at the maximum cycle temperature. The efficiency is also limited by the overall pressure ratio that can be achieved. Cycle efficiency is highest in rocket engines (~60+%), as they can achieve extremely high combustion temperatures. Cycle efficiency in turbojet and similar is nearer to 30%, due to much lower peak cycle temperatures.

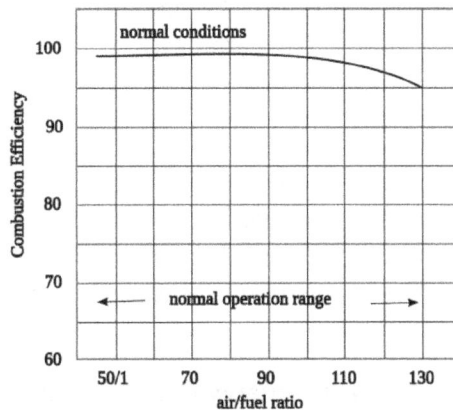

Typical combustion efficiency of an aircraft gas turbine over the operational range.

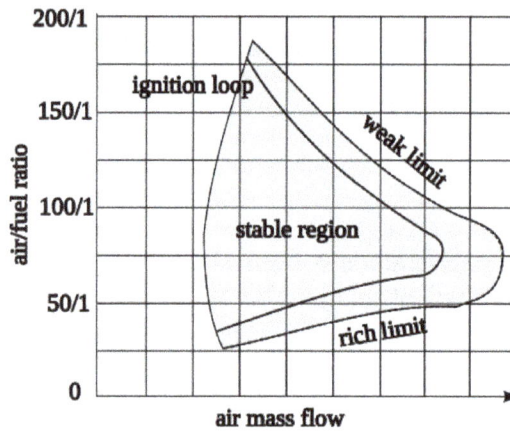

Typical combustion stability limits of an aircraft gas turbine.

The combustion efficiency of most aircraft gas turbine engines at sea level takeoff conditions is almost 100%. It decreases nonlinearly to 98% at altitude cruise conditions. Air-fuel ratio ranges from 50:1 to 130:1. For any type of combustion chamber there is a *rich* and *weak limit* to the air-fuel ratio, beyond which the flame is extinguished. The range of air-fuel ratio between the rich and weak limits is reduced with an increase of air velocity. If the increasing air mass flow reduces the fuel ratio below certain value, flame extinction occurs.

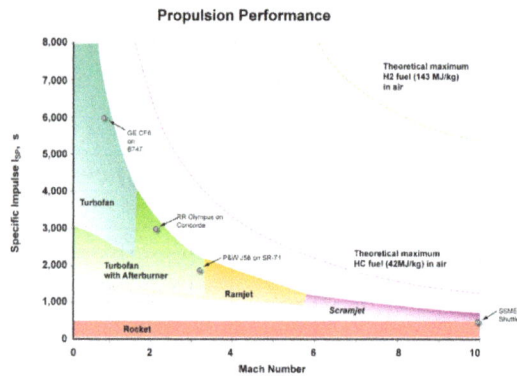

Specific impulse as a function of speed for different jet types with kerosene fuel (hydrogen I_{sp} would be about twice as high). Although efficiency plummets with speed, greater distances are covered. Efficiency per unit distance (per km or mile) is roughly independent of speed for jet engines as a group; however, airframes become inefficient at supersonic speeds.

Consumption of Fuel or Propellant

A closely related (but different) concept to energy efficiency is the rate of consumption of propellant mass. Propellant consumption in jet engines is measured by Specific Fuel Consumption, Specific impulse or Effective exhaust velocity. They all measure the same thing. Specific impulse and effective exhaust velocity are strictly proportional, whereas specific fuel consumption is inversely proportional to the others.

For airbreathing engines such as turbojets, energy efficiency and propellant (fuel) efficiency are much the same thing, since the propellant is a fuel and the source of energy. In rocketry, the propellant is also the exhaust, and this means that a high energy propellant gives better propellant efficiency but can in some cases actually give *lower* energy efficiency.

It can be seen in the table (just below) that the subsonic turbofans such as General Electric's CF6 turbofan use a lot less fuel to generate thrust for a second than did the Concorde's Rolls-Royce/Snecma Olympus 593 turbojet. However, since energy is force times distance and the distance per second was greater for Concorde, the actual power generated by the engine for the same amount of fuel was higher for Concorde at Mach 2 than the CF6. Thus, the Concorde's engines were more efficient in terms of energy per mile.

Specific fuel consumption (SFC), specific impulse, and effective exhaust velocity numbers for various rocket and jet engines.					
Engine type	Scenario	SFC in lb/(lbf·h)	SFC in g/(kN·s)	Specific impulse (s)	Effective exhaust velocity (m/s)
NK-33 rocket engine	Vacuum	10.9	308	331	3250
SSME rocket engine	Space shuttle vacuum	7.95	225	453	4440
Ramjet	Mach 1	4.5	130	800	7800
J-58 turbojet	SR-71 at Mach 3.2 (Wet)	1.9	54	1900	19000
Eurojet EJ200	Reheat	1.7	47	2200	21000
Rolls-Royce/Snecma Olympus 593 turbojet	Concorde Mach 2 cruise (Dry)	1.195	33.8	3010	29500
CF6-80C2B1F turbofan	Boeing 747-400 cruise	0.605	17.1	5950	58400
General Electric CF6 turbofan	Sea level	0.307	8.7	11700	115000

Thrust-to-Weight Ratio

The thrust-to-weight ratio of jet engines with similar configurations varies with scale, but is mostly a function of engine construction technology. For a given engine, the lighter the engine, the better the thrust-to-weight is, the less fuel is used to compensate for drag due to the lift needed to carry the engine weight, or to accelerate the mass of the engine.

As can be seen in the following table, rocket engines generally achieve much higher thrust-to-weight ratios than duct engines such as turbojet and turbofan engines. This is primarily because rockets almost universally use dense liquid or solid reaction mass which gives a much smaller volume and hence the pressurisation system that supplies the nozzle is much smaller and lighter for the same performance. Duct engines have to deal with air which is two to three orders of magnitude less dense and this gives pressures over much larger areas, which in turn results in more engineering materials being needed to hold the engine together and for the air compressor.

Jet or rocket engine	Mass (kg)	Mass (lb)	Thrust (kN)	Thrust (lbf)	Thrust-to-weight ratio
RD-0410 nuclear rocket engine	2,000	4,400	35.2	7,900	1.8

Jet or rocket engine	Mass (kg)	Mass (lb)	Thrust (kN)	Thrust (lbf)	Thrust-to-weight ratio
J58 jet engine (SR-71 Blackbird)	2,722	6,001	150	34,000	5.2
Rolls-Royce/Snecma Olympus 593 turbojet with reheat (Concorde)	3,175	7,000	169.2	38,000	5.4
Pratt & Whitney F119	1,800	3,900	91	20,500	7.95
RD-0750 rocket engine, three-propellant mode	4,621	10,188	1,413	318,000	31.2
RD-0146 rocket engine	260	570	98	22,000	38.4
SSME rocket engine (Space Shuttle)	3,177	7,004	2,278	512,000	73.1
RD-180 rocket engine	5,393	11,890	4,152	933,000	78.5
RD-170 rocket engine	9,750	21,500	7,887	1,773,000	82.5
F-1 (Saturn V first stage)	8,391	18,499	7,740.5	1,740,100	94.1
NK-33 rocket engine	1,222	2,694	1,638	368,000	136.7
Merlin 1D rocket engine, full-thrust version	467	1,030	825	185,000	180.1

Rocket thrusts are vacuum thrusts unless otherwise noted

Comparison of Types

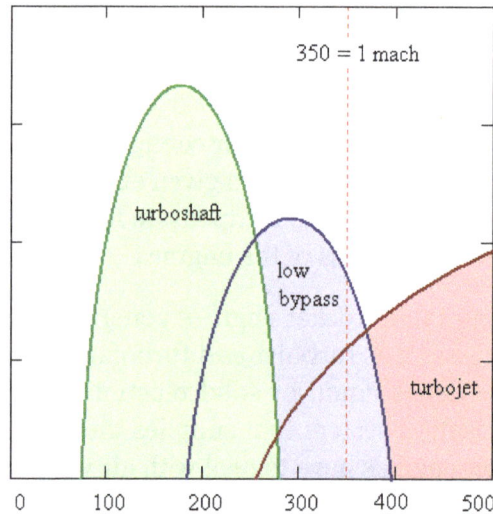

Comparative suitability for (left to right) turboshaft, low bypass and turbojet to fly at 10 km altitude in various speeds. Horizontal axis - speed, m/s. Vertical axis displays engine efficiency.

Propeller engines handle larger air mass flows, and give them smaller acceleration, than jet engines. Since the increase in air speed is small, at high flight speeds the thrust available to propeller-driven airplanes is small. However, at low speeds, these engines benefit from relatively high propulsive efficiency.

On the other hand, turbojets accelerate a much smaller mass flow of intake air and burned fuel, but they then reject it at very high speed. When a de Laval nozzle is used to accelerate a hot engine exhaust, the outlet velocity may be locally supersonic. Turbojets are particularly suitable for aircraft travelling at very high speeds.

Turbofans have a mixed exhaust consisting of the bypass air and the hot combustion product gas from the core engine. The amount of air that bypasses the core engine compared to the amount flowing into the engine determines what is called a turbofan's bypass ratio (BPR).

While a turbojet engine uses all of the engine's output to produce thrust in the form of a hot high-velocity exhaust gas jet, a turbofan's cool low-velocity bypass air yields between 30 percent and 70 percent of the total thrust produced by a turbofan system.

The net thrust (F_N) generated by a turbofan is:

$$F_N = \dot{m}_e v_e - \dot{m}_o v_o + BPR(\dot{m}_c v_f)$$

where:

$\dot{m}_e$	= the mass rate of hot combustion exhaust flow from the core engine
$\dot{m}_o$	= the mass rate of total air flow entering the turbofan = $\dot{m}_c + \dot{m}_f$
$\dot{m}_c$	= the mass rate of intake air that flows to the core engine
$\dot{m}_f$	= the mass rate of intake air that bypasses the core engine
v_f	= the velocity of the air flow bypassed around the core engine
v_e	= the velocity of the hot exhaust gas from the core engine
v_o	= the velocity of the total air intake = the true airspeed of the aircraft
BPR	= Bypass Ratio

Rocket engines have extremely high exhaust velocity and thus are best suited for high speeds (hypersonic) and great altitudes. At any given throttle, the thrust and efficiency of a rocket motor improves slightly with increasing altitude (because the back-pressure falls thus increasing net thrust at the nozzle exit plane), whereas with a turbojet (or turbofan) the falling density of the air entering the intake (and the hot gases leaving the nozzle) causes the net thrust to decrease with increasing altitude. Rocket engines are more efficient than even scramjets above roughly Mach 15.

Altitude and Speed

With the exception of scramjets, jet engines, deprived of their inlet systems can only accept air at around half the speed of sound. The inlet system's job for transonic and supersonic aircraft is to slow the air and perform some of the compression.

The limit on maximum altitude for engines is set by flammability- at very high altitudes the air becomes too thin to burn, or after compression, too hot. For turbojet engines altitudes of about 40 km appear to be possible, whereas for ramjet engines 55 km may be achievable. Scramjets may theoretically manage 75 km. Rocket engines of course have no upper limit.

At more modest altitudes, flying faster compresses the air at the front of the engine, and this greatly heats the air. The upper limit is usually thought to be about Mach 5-8, as above about Mach 5.5, the atmospheric nitrogen tends to react due to the high temperatures at the inlet and this consumes significant energy. The exception to this is scramjets which may be able to achieve about Mach 15 or more, as they avoid slowing the air, and rockets again have no particular speed limit.

Noise

The noise emitted by a jet engine has many sources. These include, in the case of gas turbine engines, the fan, compressor, combustor, turbine and propelling jet/s.

The propelling jet produces jet noise which is caused by the violent mixing action of the high speed jet with the surrounding air. In the subsonic case the noise is produced by eddies and in the supersonic case by Mach waves. The sound power radiated from a jet varies with the jet velocity raised to the eighth power for velocities up to 2,000 ft/sec and varies with the velocity cubed above 2,000 ft/sec. Thus, the lower speed exhaust jets emitted from engines such as high bypass turbofans are the quietest, whereas the fastest jets, such as rockets, turbojets, and ramjets, are the loudest. For commercial jet aircraft the jet noise has reduced from the turbojet through bypass engines to turbofans as a result of a progressive reduction in propelling jet velocities. For example, the JT8D, a bypass engine, has a jet velocity of 1450 ft/sec whereas the JT9D, a turbofan, has jet velocities of 885 ft/sec (cold) and 1190 ft/sec (hot).

The advent of the turbofan replaced the very distinctive jet noise with another sound known as "buzz saw" noise. The origin is the shockwaves originating at the supersonic fan blades at takeoff thrust.

Types of Jet Engine

Turbojet

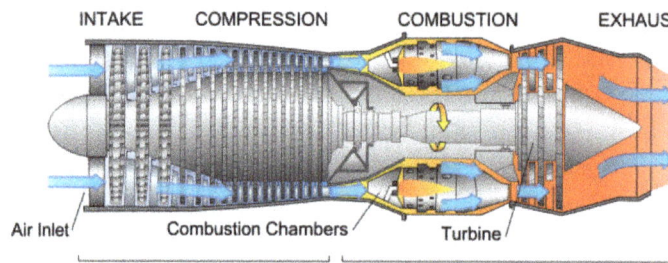

Diagram of a typical gas turbine jet engine

The turbojet is an airbreathing jet engine, usually used in aircraft. It consists of a gas turbine with a propelling nozzle. The gas turbine has an air inlet, a compressor, a combustion chamber, and a turbine (that drives the compressor). The compressed air from the compressor is heated by the fuel in the combustion chamber and then allowed to expand through the turbine. The turbine exhaust is then expanded in the propelling nozzle where it is accelerated to high speed to provide thrust. Two engineers, Frank Whittle in the United Kingdom and Hans von Ohain in Germany, developed the concept independently into practical engines during the late 1930s.

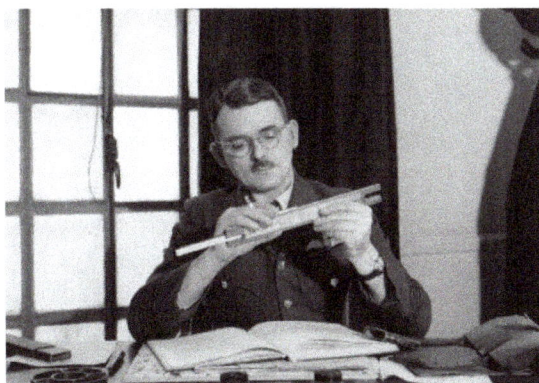
Frank Whittle

Turbojets have been replaced in slower aircraft by turboprops because they have better range-specific fuel consumption. At medium speeds, where the propeller is no longer efficient, turboprops have been replaced by turbofans. The turbofan is quieter and has better range-specific fuel consumption than the turbojet. Turbojets are still common in medium range cruise missiles, due to their high exhaust speed, small frontal area, and relative simplicity.

Hans von Ohain

Turbojets have poor efficiency at low vehicle speeds, which limits their usefulness in vehicles other than aircraft. Turbojet engines have been used in isolated cases to power vehicles other than aircraft, typically for attempts on land speed records. Where vehicles are 'turbine powered' this is more commonly by use of a turboshaft engine, a development of the gas turbine engine where an additional turbine is used to drive a rotating output shaft. These are common in helicopters and hovercraft. Turbojets have also been used experimentally to clear snow from switches in railyards.

History

The first patent for using a gas turbine to power an aircraft was filed in 1921 by Frenchman Maxime Guillaume. His engine was to be an axial-flow turbojet, but was never constructed, as it would have required considerable advances over the state of the art in compressors.

Zu der Patentschrift 554906
Kl. 46d Gr. 17

Abb. 1

Abb. 2

Abb. 3

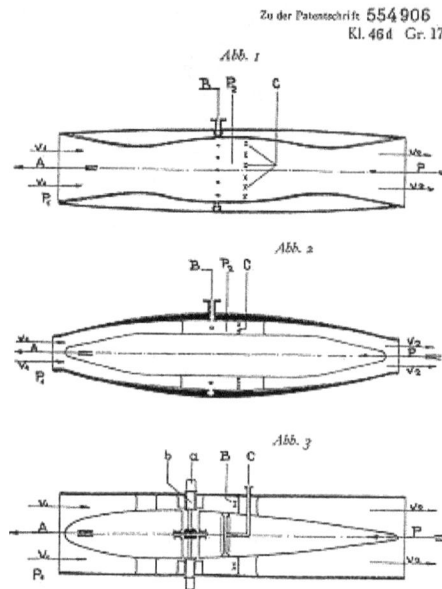

Albert Fonó's German patent for jet engines (January 1928). The third illustration is a turbojet

Heinkel He 178, the world's first aircraft to fly purely on turbojet power, using an HeS 3 engine

Practical axial compressors were made possible by ideas from A.A.Griffith in a seminal paper in 1926 ("An Aerodynamic Theory of Turbine Design").

The centrifugal-flow turbojet was first patented in 1930 by Frank Whittle of the Royal Air Force, and in Germany, Hans von Ohain patented a similar engine in 1935.

The first turbojet to run was the Power Jets WU which ran on 12 April 1937.

On 27 August 1939 the Heinkel He 178 became the world's first aircraft to fly under turbojet power with test-pilot Erich Warsitz at the controls, thus becoming the first practical jet plane. The first two operational turbojet aircraft, the Messerschmitt Me 262 and then the Gloster Meteor entered service in 1944 towards the end of World War II.

Air is drawn into the rotating compressor via the intake and is compressed to a higher pressure before entering the combustion chamber. Fuel is mixed with the compressed air and burns in the combustor. The combustion products leave the combustor and expand through the turbine where power is extracted to drive the compressor. The turbine exit gases still contain considerable energy that is converted in the propelling nozzle to a high speed jet.

The first jet engines were turbojets, with either a centrifugal compressor (as in the Heinkel HeS 3), or Axial compressors (as in the Junkers Jumo 004) which gave a smaller diameter, although longer, engine. By replacing the propeller used on piston engines with a high speed jet of exhaust higher aircraft speeds were attainable.

One of the last applications for a turbojet engine was the Concorde which used the Olympus 593 engine. At the time of its design the turbojet was still seen as the optimum for cruising at twice the speed of sound despite the advantage of turbofans for lower speeds. For the Concorde less fuel was required to produce a given thrust for a mile at Mach 2.0 than a modern high-bypass turbofan such as General Electric CF6 at its Mach 0.86 optimum speed.

Turbojet engines had a significant impact on commercial aviation. Aside from giving faster flight speeds turbojets had greater reliability than piston engines, with some models demonstrating dispatch reliability rating in excess of 99.9%. Pre-jet commercial aircraft were designed with as many as 4 engines in part because of concerns over in-flight failures. Overseas flight paths were plotted to keep planes within an hour of a landing field, lengthening flights. The increase in reliability that came with the turbojet enabled three and two-engine designs, and more direct long-distance flights.

High-temperature alloys were a reverse salient, a key technology that dragged progress on jet engines. Non-UK jet engines built in the 1930s and 1940s had to be overhauled every 10 or 20 hours due to creep failure and other types of damage to blades. British engines however utilised Nimonic alloys which allowed extended use without overhaul, engines such as the Rolls-Royce Welland and Rolls-Royce Derwent, and by 1949 the de Havilland Goblin, being type tested for 500 hours without maintenance. It was not until the 1950s that superalloy technology allowed other countries to produce economically practical engines.

Early Designs

Early German turbojets had severe limitations on the amount of running they could do due to the lack of suitable high temperature materials for the turbines. British engines such as the Rolls-Royce Welland used better materials giving improved durability. The Welland was type certificated for 80 hours initially, later extended to 150 hours between overhauls, as a result of an extended 500 hour run being achieved in tests. A few of the original fighters still exist with their original engines, but many have been re-engined with more modern engines with greater fuel efficiency and a longer TBO (such as the reproduction Me-262 powered by General Electric J85s).

J85-GE-17A turbojet engine from General Electric (1970)

General Electric in the United States was in a good position to enter the jet engine business due to its experience with the high temperature materials used in their turbosuperchargers during World War II.

Water injection was a common method used to increase thrust, usually during takeoff, in early turbojets that were thrust-limited by their allowable turbine entry temperature. The water increased thrust at the temperature limit but prevented complete combustion often leaving a very visible smoke trail.

Allowable turbine entry temperatures have increased steadily over time both with the introduction of superior alloys, and coatings, and with the introduction and progressive effectiveness of blade cooling designs. On early engines the turbine temperature limit had to be monitored, and avoided, by the pilot, typically during starting and at maximum thrust settings. Automatic temperature limiting was introduced to reduce pilot workload and reduce the liklehood of turbine damage due to overtemperature.

Design

TF33-P-7 engine of a C-141B showing zero-length inlet

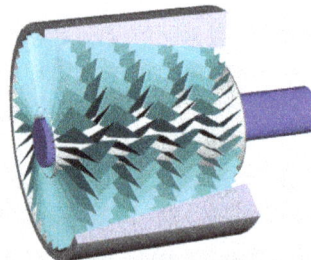

An animation of an axial compressor. The stationary blades are the stators

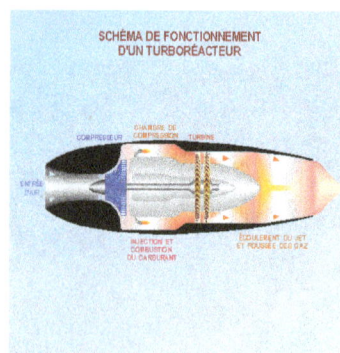

Turbojet animation

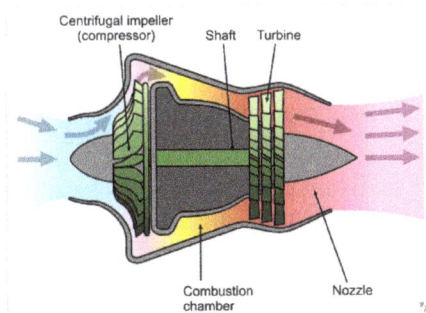

Schematic diagram showing the operation of a centrifugal flow turbojet engine. The compressor is driven via the turbine stage and throws the air outwards, requiring it to be redirected parallel to the axis of thrust

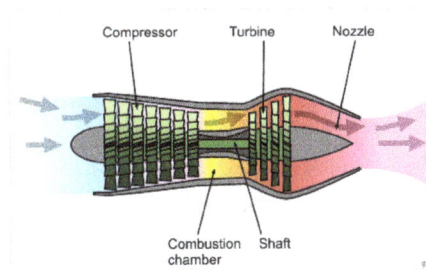

Schematic diagram showing the operation of an axial flow turbojet engine. Here, the compressor is again driven by the turbine, but the air flow remains parallel to the axis of thrust

Air Intake

An intake, or tube, is needed in front of the compressor to help direct the incoming air smoothly into the moving compressor blades. Older engines had stationary vanes in front of the moving blades. These vanes also helped to direct the air onto the blades. The air flowing into a turbojet engine is always subsonic, regardless of the speed of the aircraft itself.

The intake has to supply air to the engine with an acceptably small variation in pressure (known as distortion) and having lost as little energy as possible on the way (known as pressure recovery). The ram pressure rise in the intake is the inlets contribution to the propulsion system overall pressure ratio and thermal efficiency.

The intake gains prominence at high speeds when it transmits more thrust to the airframe than the engine does. Well-known examples are the Concorde and Lockheed SR-71 Blackbird propulsion systems where the intake and engine contributions to the total powerplant were 63%/8% at Mach 2 and 54%/17% at Mach 3+.

Intakes have ranged from 'zero-length' on the Pratt & Whitney TF33 installation in the Lockheed C-141 Starlifter to the twin, 65 feet-long, intakes on the North American XB-70 Valkyrie each feeding three engines with an intake airflow of about 800 lb/sec.

Compressor

The compressor is driven by the turbine. It rotates at high speed, adding energy to the airflow and at the same time squeezing (compressing) it into a smaller space. Compressing the air increases its pressure and temperature. The smaller the compressor the faster it turns. At the large end of the range the GE-90-115 fan rotates at about 2,500 RPM while a small helicopter engine compressor rotates at about 50,000 RPM.

Turbojets supply bleed air from the compressor to the aircraft for the Environmental Control System, anti-icing and fuel tank pressurization, for example. The engine itself needs air at various pressures and flow rates to keep it running. This air comes from the compressor and without it the turbines would overheat, the lubricating oil would leak from the bearing cavities, the rotor thrust bearings would skid or be overloaded and ice would form on the nose cone. The air from the compressor is called secondary air and is used for turbine cooling, bearing cavity sealing, anti-icing and ensuring that the rotor axial load on its thrust bearing will not wear it out prematurely. Supplying bleed air to the aircraft decreases the efficiency of the engine because it has been compressed but then does not contribute to producing thrust. Bleed air for aircraft services is no longer needed on the turbofan-powered Boeing 787.

Compressor types used in turbojets were typically axial or centrifugal.

Early turbojet compressors had low pressure ratios up to about 5:1. Aerodynamic improvements including splitting the compressor into two separately rotating parts, incorporating variable blade angles for entry guide vanes and stators and bleeding air from the compressor enabled later turbojets to have overall pressure ratios of 15:1 or more. For comparison, modern civil turbofan engines have overall pressure ratios of 44:1 or more.

After leaving the compressor, the air enters the combustion chamber.

Combustion Chamber

The burning process in the combustor is significantly different from that in a piston engine. In a piston engine the burning gases are confined to a small volume and, as the fuel burns, the pressure increases. In a turbojet the air and fuel mixture burn in the combustor and pass through to the turbine in a continuous flowing process with no pressure build-up. Instead there is a small pressure loss in the combustor.

The fuel-air mixture can only burn in slow moving air so an area of reverse flow is maintained by the fuel nozzles for the approximately stoichiometric burning in the primary zone. Further compressor air is introduced which completes the combustion process and reduces the temperature of the combustion products to a level which the turbine can accept. Less than 25% of the air is typically used for combustion, as an overall lean mixture is required to keep within the turbine temperature limits.

Turbine

Hot gases leaving the combustor expand through the turbine. Typical materials for turbines include inconel and Nimonic. The hottest turbine vanes and blades in an engine have internal cooling passages. Air from the compressor is passed through these to keep the metal temperature within limits. The remaining stages don't need cooling.

In the first stage the turbine is largely an impulse turbine (similar to a pelton wheel) and rotates because of the impact of the hot gas stream. Later stages are convergent ducts that accelerate the gas. Energy is transferred into the shaft through momentum exchange in the opposite way to energy transfer in the compressor. The power developed by the turbine drives the compressor as well as accessories, like fuel, oil, and hydraulic pumps that are driven by the accessory gearbox.

Nozzle

After the turbine, the gases expand through the exhaust nozzle producing a high velocity jet. In a convergent nozzle, the ducting narrows progressively to a throat. The nozzle pressure ratio on a turbojet is high enough at higher thrust settings to cause the nozzle to choke.

If, however, a convergent-divergent de Laval nozzle is fitted, the divergent (increasing flow area) section allows the gases to reach supersonic velocity within the divergent section. Additional thrust is generated by the higher resulting exhaust velocity.

Thrust Augmentation

Thrust was most commonly increased in turbojets with water/methanol injection or afterburning. Some engines used both at the same time.

Liquid injection was tested on the Power Jets W.1 in 1941 initially using ammonia before changing to water and then water/methanol. A system to trial the technique in the Gloster E.28/39 was devised but never fitted.

Afterburner

An afterburner or "reheat jetpipe" is a combustion chamber added to reheat the turbine exhaust gases. The fuel consumption is very high, typically four times that of the main engine. Afterburners are used almost exclusively on supersonic aircraft, most being military aircraft. Two supersonic airliners, Concorde and the Tu-144, also used afterburners as does Scaled Composites White Knight, a carrier aircraft for the experimental SpaceShipOne suborbital spacecraft.

Reheat was flight-trialled in 1944 on the W.2/700 engines in a Gloster Meteor I.

Net Thrust

The net thrust F_N of a turbojet is given by:

$$F_N = (\dot{m}_{air} + \dot{m}_f)V_j - \dot{m}_{air}V$$

where:

$\dot{m}_{air}$	is the rate of flow of air through the engine
$\dot{m}_f$	is the rate of flow of fuel entering the engine
V_j	is the speed of the jet (the exhaust plume) and is assumed to be less than sonic velocity

V	is the <u>true airspeed</u> of the aircraft
$(\dot{m}_{air} + \dot{m}_f)V_j$	represents the nozzle gross thrust
$\dot{m}_{air}V$	represents the ram drag of the intake

If the speed of the jet is equal to sonic velocity the nozzle is said to be choked. If the nozzle is *choked* the pressure at the nozzle exit plane is greater than atmospheric pressure, and extra terms must be added to the above equation to account for the *pressure thrust*.

The rate of flow of fuel entering the engine is very small compared with the rate of flow of air. If the contribution of fuel to the nozzle gross thrust is ignored, the net thrust is:

$$F_N = \dot{m}_{air}(V_j - V)$$

The speed of the jet V_j must exceed the true airspeed of the aircraft V if there is to be a net forward thrust on the airframe. The speed V_j can be calculated thermodynamically based on adiabatic expansion.

Cycle Improvements

The operation of a turbojet is modelled approximately by the Brayton Cycle.

The efficiency of a gas turbine is increased by raising the overall pressure ratio, requiring higher temperature compressor materials, and raising the turbine entry temperature, requiring better turbine materials and/or improved vane/blade cooling. It is also increased by reducing the losses as the flow progresses from the intake to the propelling nozzle. These losses are quantified by compressor and turbine efficiencies and ducting pressure losses. When used in a turbojet application, where the output from the gas turbine is used in a propelling nozzle, raising the turbine temperature increases the jet velocity. This reduces the propulsive efficiency giving a loss in overall efficiency, as reflected by the higher fuel consumption, or SFC.

References

Springer, Edwin H. (2001). Constructing A Turbocharger Turbojet Engine. Turbojet Technologies.

Turbofan

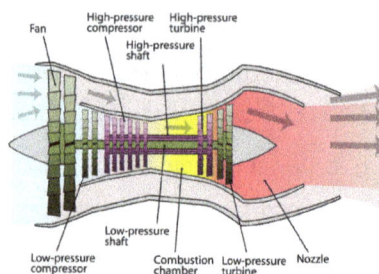

Schematic diagram of a high-bypass turbofan engine

Rolls-Royce Trent 1000 turbofan powering a Boeing 787 Dreamliner testflight

Engine Alliance GP7000 turbofan awaiting installation on an Airbus A380 under construction

The turbofan or fanjet is a type of airbreathing jet engine that is widely used in aircraft propulsion. The word "turbofan" is a portmanteau of "turbine" and "fan": the *turbo* portion refers to a gas turbine engine which achieves mechanical energy from combustion, and the *fan*, a ducted fan that uses the mechanical energy from the gas turbine to accelerate air rearwards. Thus, whereas all the air taken in by a turbojet passes through the turbine (through the combustion chamber), in a turbofan some of that air bypasses the turbine. A turbofan thus can be thought of as a turbojet being used to drive a ducted fan, with both of those contributing to the thrust. The ratio of the mass-flow of air bypassing the engine core compared to the mass-flow of air passing through the core is referred to as the bypass ratio. The engine produces thrust through a combination of these two portions working in concert; engines that use more jet thrust relative to fan thrust are known as *low bypass turbofans*, conversely those that have considerably more fan thrust than jet thrust are known as *high bypass*. Most commercial aviation jet engines in use today are of the high-by-pass type, and most modern military fighter engines are low-bypass. Afterburners are not used on high-bypass turbofan engines but may be used on either low-bypass turbofan or turbojet engines.

Most of the air flow through a high-bypass turbofan is low-velocity bypass flow: even when combined with the much higher velocity engine exhaust, the average exhaust velocity is considerably lower than in a pure turbojet. Turbojet engine noise is predominately jet noise from the high exhaust velocity, therefore turbofan engines are significantly quieter than a pure-jet of the same thrust with jet noise no longer the predominant source. Other noise sources are the fan, compressor and turbine. Jet noise is reduced with chevrons, sawtooth patterns on the exhaust nozzles, on the Rolls-Royce Trent 1000 and General Electric GEnx engines used on the Boeing 787.

Since the efficiency of propulsion is a function of the relative airspeed of the exhaust to the surrounding air, propellers are most efficient for low speed, pure jets for high speeds, and ducted fans in the middle. Turbofans are thus the most efficient engines in the range of speeds from about 500 to 1,000 km/h (310 to 620 mph), the speed at which most commercial aircraft operate. Turbofans retain an efficiency edge over pure jets at low supersonic speeds up to roughly Mach 1.6.

Modern turbofans have either a large single-stage fan or a smaller fan with several stages. An early configuration combined a low-pressure turbine and fan in a single rear-mounted unit.

Early Turbofans

Rolls-Royce Conway low bypass turbofan from a Boeing 707. The bypass air exits from the fins whilst the exhaust from the core exits from the central nozzle. This fluted jetpipe design is a noise-reducing method devised by Frederick Greatorex at Rolls-Royce

Early turbojet engines were not very fuel-efficient as their overall pressure ratio and turbine inlet temperature were severely limited by the technology available at the time. In 1939-1941 Soviet designer Arkhip Lyulka elaborated the design for the world's first turbofan engine, and acquired a patent for this new invention on April 22, 1941. Although several prototypes were built and ready for testing, Lyulka was in 1941 forced to abandon his research and evacuate to the Urals following the Nazi invasion of the Soviet Union. So the first turbofan to run was apparently the German Daimler-Benz DB 670 (designated as the 109-007 by the RLM) with a first run date of 27 May 1943. Turbomachinery testing, using an electric motor, had started on 1 April 1943. The engine was abandoned later while the war went on and problems could not be solved. The British wartime Metrovick F.2 axial flow jet was given a fan, as the Metrovick F.3 in 1943, to create the first British turbofan.

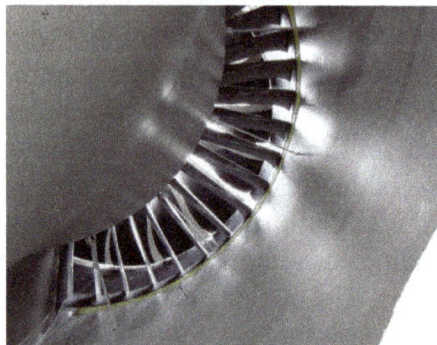

General Electric GEnx-2B turbofan engine from a Boeing 747-8. View into the outer (propelling or "cold") nozzle.

Improved materials, and the introduction of twin compressors such as in the Bristol Olympus and Pratt & Whitney JT3C engines, increased the overall pressure ratio and thus the thermodynamic efficiency of engines, but they also led to a poor propulsive efficiency, as pure turbojets have a high specific thrust/high velocity exhaust better suited to supersonic flight.

The original low-bypass turbofan engines were designed to improve propulsive efficiency by reducing the exhaust velocity to a value closer to that of the aircraft. The Rolls-Royce Conway, the world's first production turbofan, had a bypass ratio of 0.3, similar to the modern General Electric F404 fighter engine. Civilian turbofan engines of the 1960s, such as the Pratt & Whitney JT8D and the Rolls-Royce Spey had bypass ratios closer to 1, and were similar to their military equivalents.

The first General Electric turbofan was the aft-fan CJ805-23 based on the CJ805-3 turbojet. It was followed by the aft-fan General Electric CF700 engine with a 2.0 bypass ratio. This was derived from the General Electric J85/CJ610 turbojet (2,850 lbf or 12,650 N) to power the larger Rockwell Sabreliner 75/80 model aircraft, as well as the Dassault Falcon 20 with about a 50% increase in thrust (4,200 lbf or 18,700 N). The CF700 was the first small turbofan in the world to be certified by the Federal Aviation Administration (FAA). There were at one time over 400 CF700 aircraft in operation around the world, with an experience base of over 10 million service hours. The CF700 turbofan engine was also used to train Moon-bound astronauts in Project Apollo as the powerplant for the Lunar Landing Research Vehicle.

Low-Bypass Turbofan

A high specific thrust/low bypass ratio turbofan normally has a multi-stage fan, developing a relatively high pressure ratio and, thus, yielding a high (mixed or cold) exhaust velocity. The core airflow needs to be large enough to give sufficient core power to drive the fan. A smaller core flow/ higher bypass ratio cycle can be achieved by raising the (HP) turbine rotor inlet temperature.

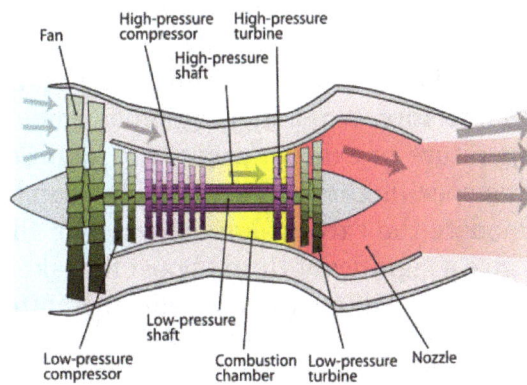

Schematic diagram illustrating a 2-spool, low-bypass turbofan engine with a mixed exhaust, showing the low-pressure (green) and high-pressure (purple) spools. The fan (and booster stages) are driven by the low-pressure turbine, whereas the high-pressure compressor is powered by the high-pressure turbine

To illustrate one aspect of how a turbofan differs from a turbojet they may be compared, as in a re-engining assessment, at the same airflow (to keep a common intake for example) and the same net thrust (i.e. same specific thrust). A bypass flow can only be added to the turbojet if the turbine inlet temperature is allowed to increase to compensate for the smaller core flow. Improvements in turbine cooling/material technology would allow the use of a higher turbine inlet temperature despite an increase in cooling air temperature which would result from any overall pressure-ratio increase.

The resulting turbofan, with reasonable efficiencies and duct loss for the added components, would probably operate at a higher nozzle pressure ratio than the turbojet, but with a lower exhaust temperature to retain net thrust. Since the temperature rise across the whole engine (intake to nozzle) would be lower, the (dry power) fuel flow would also be reduced, resulting in a better specific fuel consumption (SFC).

Some low-bypass ratio military turbofans (e.g. F404) have variable inlet guide vanes to direct air onto the first fan rotor stage. This improves the fan surge margin.

Afterburning Turbofan

Pratt & Whitney F119 afterburning turbofan on test

Since the 1970s, most jet fighter engines have been low/medium bypass turbofans with a mixed exhaust, afterburner and variable area final nozzle. An afterburner is a combustor located downstream of the turbine blades and directly upstream of the nozzle, which burns fuel from afterburner-specific fuel injectors. When lit, prodigious amounts of fuel are burnt in the afterburner, raising the temperature of exhaust gases by a significant degree, resulting in a higher exhaust velocity/engine specific thrust. The variable geometry nozzle must open to a larger throat area to accommodate the extra volume flow when the afterburner is lit. Afterburning is often designed to give a significant thrust boost for take off, transonic acceleration and combat maneuvers, but is very fuel intensive. Consequently, afterburning can only be used for short portions of a mission.

Unlike the main combustor, where the downstream turbine blades must not be damaged by high temperatures, an afterburner can operate at the ideal maximum (stoichiometric) temperature (i.e., about 2100K/3780Ra/3320F). At a fixed total applied fuel:air ratio, the total fuel flow for a given fan airflow will be the same, regardless of the dry specific thrust of the engine. However, a high specific thrust turbofan will, by definition, have a higher nozzle pressure ratio, resulting in a higher afterburning net thrust and, therefore, a lower afterburning specific fuel consumption (SFC). However, high specific thrust engines have a high dry SFC. The situation is reversed for a medium specific thrust afterburning turbofan: i.e., poor afterburning SFC/good dry SFC. The former engine is suitable for a combat aircraft which must remain in afterburning combat for a fairly long period, but only has to fight fairly close to the airfield (e.g. cross border skirmishes) The latter engine is better for an aircraft that has to fly some distance, or loiter for a long time, before going into combat. However, the pilot can only afford to stay in afterburning for a short period, before aircraft fuel reserves become dangerously low.

The first production afterburning turbofan engine was the Pratt & Whitney TF30, which initially powered the F-111 Aardvark and F-14 Tomcat. Current low-bypass military turbofans include the Pratt & Whitney F119, the Eurojet EJ200, the General Electric F110, the Klimov RD-33, and the Saturn AL-31, all of which feature a mixed exhaust, afterburner and variable area propelling nozzle.

High-Bypass Turbofan

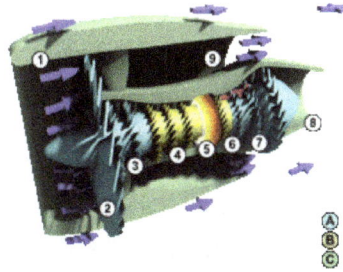

Animation of a 2-spool, high-bypass turbofan.A. Low-pressure spool

B. High-pressure spool

C. Stationary components

1. Nacelle

2. Fan

3. Low-pressure compressor

4. High-pressure compressor

5. Combustion chamber

6. High-pressure turbine

7. Low-pressure turbine

8. Core nozzle

9. Fan nozzle

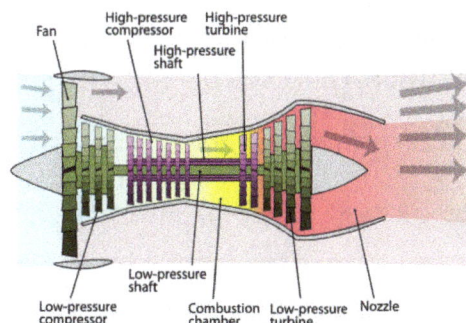

Schematic diagram illustrating a 2-spool, high-bypass turbofan engine with an unmixed exhaust. The low-pressure spool is coloured green and the high-pressure one purple. Again, the fan (and booster stages) are driven by the low-pressure turbine, but more stages are required. A mixed exhaust is often employed nowadays.

The low specific thrust/high bypass ratio turbofans used in today's civil jetliners (and some military transport aircraft) evolved from the high specific thrust/low bypass ratio turbofans used in such [production] aircraft back in the 1960s.

Low specific thrust is achieved by replacing the multi-stage fan with a single-stage unit. Unlike some military engines, modern civil turbofans do not have any stationary inlet guide vanes in front of the fan rotor. The fan is scaled to achieve the desired net thrust.

The core (or gas generator) of the engine must generate sufficient core power to at least drive the fan at its design flow and pressure ratio. Through improvements in turbine cooling/material technology, a higher (HP) turbine rotor inlet temperature can be used, thus facilitating a smaller (and lighter) core and (potentially) improving the core thermal efficiency. Reducing the core mass flow tends to increase the load on the LP turbine, so this unit may require additional stages to reduce the average stage loading and to maintain LP turbine efficiency. Reducing core flow also increases bypass ratio. Bypass ratios greater than 5:1 are increasingly common with the Rolls-Royce Trent XWB approaching 10:1.

Further improvements in core thermal efficiency can be achieved by raising the overall pressure ratio of the core. Improved blade aerodynamics reduces the number of extra compressor stages required. With multiple compressors (i.e., LPC, IPC, and HPC) dramatic increases in overall pressure ratio have become possible. Variable geometry (i.e., stators) enable high-pressure-ratio compressors to work surge-free at all throttle settings.

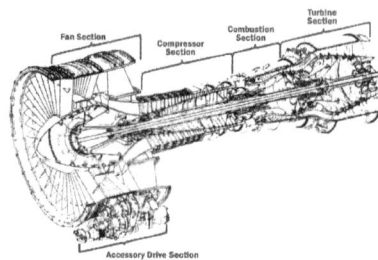

Cutaway diagram of the General Electric CF6-6 engine

The first (experimental) high bypass turbofan engine was built and run on February 13, 1964 by AVCO-Lycoming. Shortly after, the General Electric TF39 became the first production model, designed to power the Lockheed C-5 Galaxy military transport aircraft. The civil General Electric CF6 engine used a derived design. Other high-bypass turbofans are the Pratt & Whitney JT9D, the three-shaft Rolls-Royce RB211 and the CFM International CFM56; also the smaller TF34. More recent large high-bypass turbofans include the Pratt & Whitney PW4000, the three-shaft Rolls-Royce Trent, the General Electric GE90/GEnx and the GP7000, produced jointly by GE and P&W.

For reasons of fuel economy, and also of reduced noise, almost all of today's jet airliners are powered by high-bypass turbofans. Although modern combat aircraft tend to use low bypass ratio turbofans, military transport aircraft (e.g., C-17) mainly use high bypass ratio turbofans (or turboprops) for fuel efficiency.

The lower the specific thrust of a turbofan, the lower the mean jet outlet velocity, which in turn translates into a high thrust lapse rate (i.e. decreasing thrust with increasing flight speed). See technical discussion below, item 2. Consequently, an engine sized to propel an aircraft at high subsonic flight speed (e.g., Mach 0.83) generate a relatively high thrust at low flight speed, thus enhancing runway performance. Low specific thrust engines tend to have a high bypass ratio, but this is also a function of the temperature of the turbine system.

The turbofans on twin engined airliners are further more powerful to cope with losing one engine during take-off, which reduces the aircraft's net thrust by half. Modern twin engined airliners normally climb very steeply immediately after take-off. If one engine is lost, the climb-out is much shallower, but sufficient to clear obstacles in the flightpath.

The Soviet Union's engine technology was less advanced than the West's and its first wide-body aircraft, the Ilyushin Il-86, was powered by low-bypass engines. The Yakovlev Yak-42, a medium-range, rear-engined aircraft seating up to 120 passengers introduced in 1980 was the first Soviet aircraft to use high-bypass engines.

Turbofan Configurations

Turbofan engines come in a variety of engine configurations. For a given engine cycle (i.e., same airflow, bypass ratio, fan pressure ratio, overall pressure ratio and HP turbine rotor inlet temperature), the choice of turbofan configuration has little impact upon the design point performance (e.g., net thrust, SFC), as long as overall component performance is maintained. Off-design performance and stability is, however, affected by engine configuration.

As the design overall pressure ratio of an engine cycle increases, it becomes more difficult to throttle the compression system, without encountering an instability known as compressor surge. This occurs when some of the compressor aerofoils stall (like the wings of an aircraft) causing a violent change in the direction of the airflow. However, compressor stall can be avoided, at throttled conditions, by progressively:

opening interstage/intercompressor blow-off valves (inefficient), and/or

closing variable stators within the compressor

Most modern American civil turbofans employ a relatively high-pressure-ratio high-pressure (HP) compressor, with many rows of variable stators to control surge margin at part-throttle. In the three-spool RB211/Trent the core compression system is split into two, with the IP compressor, which supercharges the HP compressor, being on a different coaxial shaft and driven by a separate (IP) turbine. As the HP compressor has a modest pressure ratio it can be throttled-back surge-free, without employing variable geometry. However, because a shallow IP compressor working line is inevitable, the IPC has one stage of variable geometry on all variants except the -535, which has none.

Single-Shaft Turbofan

Although far from common, the single-shaft turbofan is probably the simplest configuration, comprising a fan and high-pressure compressor driven by a single turbine unit, all on the same shaft. The SNECMA M53, which powers Mirage fighter aircraft, is an example of a single-shaft turbofan. Despite the simplicity of the turbomachinery configuration, the M53 requires a variable area mixer to facilitate part-throttle operation.

Aft-Fan Turbofan

One of the earliest turbofans was a derivative of the General Electric J79 turbojet, known as the CJ805-23, which featured an integrated aft fan/low-pressure (LP) turbine unit located in the turbojet exhaust jetpipe. Hot gas from the turbojet turbine exhaust expanded through the LP turbine, the fan blades being a radial extension of the turbine blades. This aft-fan configuration was later exploited in the General Electric GE-36 UDF (propfan) demonstrator of the early 80s. One of the problems with the aft fan configuration is hot gas leakage from the LP turbine to the fan.

Basic Two Spool

Many turbofans have the basic two-spool configuration where both the fan and LP turbine (i.e., LP spool) are mounted on a second (LP) shaft, running concentrically with the HP spool (i.e., HP compressor driven by HP turbine). The BR710 is typical of this configuration. At the smaller thrust sizes, instead of all-axial blading, the HP compressor configuration may be axial-centrifugal (e.g., General Electric CFE738), double-centrifugal or even diagonal/centrifugal (e.g., Pratt & Whitney Canada PW600).

Boosted Two Spool

Higher overall pressure ratios can be achieved by either raising the HP compressor pressure ratio or adding an intermediate-pressure (IP) compressor between the fan and HP compressor, to supercharge or boost the latter unit helping to raise the overall pressure ratio of the engine cycle to the very high levels employed today (i.e., greater than 40:1, typically). All of the large American turbofans (e.g., General Electric CF6, GE90 and GEnx plus Pratt & Whitney JT9D and PW4000) feature an IP compressor mounted on the LP shaft and driven, like the fan, by the LP turbine, the mechanical speed of which is dictated by the tip speed and diameter of the fan. The Rolls-Royce BR715 is a non-American example of this. The high bypass ratios (i.e., fan duct flow/core flow) used in modern civil turbofans tends to reduce the relative diameter of the attached IP compressor, causing its mean tip speed to decrease. Consequently, more IPC stages are required to develop the necessary IPC pressure rise.

Three Spool

Rolls-Royce chose a three spool configuration for their large civil turbofans (i.e., the RB211 and Trent families), where the intermediate pressure (IP) compressor is mounted on a separate (IP) shaft, running concentrically with the LP and HP shafts, and is driven by a separate IP turbine. The first three spool engine was the earlier Rolls-Royce RB.203 Trent of 1967.

Ivchenko Design Bureau chose the same configuration for their Lotarev D-36 engine, followed by Lotarev/Progress D-18T and Progress D-436.

The Turbo-Union RB199 military turbofan also has a three spool configuration, as do the military Kuznetsov NK-25 and NK-321.

Geared Fan

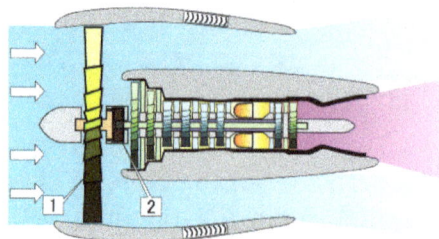

Geared turbofan

As bypass ratio increases, the mean radius ratio of the fan and low-pressure turbine (LPT) increases. Consequently, if the fan is to rotate at its optimum blade speed the LPT blading will spin slowly, so additional LPT stages will be required, to extract sufficient energy to drive the fan. Introducing a (planetary) reduction gearbox, with a suitable gear ratio, between the LP shaft and the fan enables both the fan and LP turbine to operate at their optimum speeds. Typical of this configuration are the long-established Honeywell TFE731, the Honeywell ALF 502/507, and the recent Pratt & Whitney PW1000G.

Military Turbofans

Ducting on a Dassault/Dornier Alpha Jet — At subsonic speeds, the increasing diameter of the inlet duct slows incoming air, causing its static pressure to increase.

Most of the configurations discussed above are used in civilian turbofans, while modern military turbofans (e.g., SNECMA M88) are usually basic two-spool.

High-Pressure Turbine

Most civil turbofans use a high-efficiency, 2-stage HP turbine to drive the HP compressor. The CFM56 uses an alternative approach: a single-stage, high-work unit. While this approach is probably less efficient, there are savings on cooling air, weight and cost.

In the RB211 and Trent 3-spool engine series, the HP compressor pressure ratio is modest so only a single HP stage is required. Rather than adding stage/s to the LP turbine to drive the higher pressure ratio IP (intermediate pressure) compressor, Rolls-Royce mounts it on a separate shaft and drives it with an IP turbine.

Because the HP compressor pressure ratio is modest, modern military turbofans tend to use a single-stage HP turbine.

Low-Pressure Turbine

Modern civil turbofans have multi-stage LP turbines (e.g., 3, 4, 5, 6, 7). The number of stages required depends on the engine cycle bypass ratio and how much supercharging (i.e., IP compression) is on the LP shaft, behind the fan. A geared fan may reduce the number of required LPT stages in some applications. Because of the much lower bypass ratios employed, military turbofans only require one or two LP turbine stages.

Cycle Improvements

Consider a mixed turbofan with a fixed bypass ratio and airflow. Increasing the overall pressure ratio of the compression system raises the combustor entry temperature. Therefore, at a fixed fuel flow there is an increase in (HP) turbine rotor inlet temperature. Although the higher temperature rise across the compression system implies a larger temperature drop over the turbine system, the mixed nozzle temperature is unaffected, because the same amount of heat is being added to the system. There is, however, a rise in nozzle pressure, because overall pressure ratio increases faster than the turbine expansion ratio, causing an increase in the hot mixer entry pressure. Consequently, net thrust increases, whilst specific fuel consumption (fuel flow/net thrust) decreases. A similar trend occurs with unmixed turbofans.

So turbofans can be made more fuel efficient by raising overall pressure ratio and turbine rotor inlet temperature in unison. However, better turbine materials and/or improved vane/blade cooling are required to cope with increases in both turbine rotor inlet temperature and compressor delivery temperature. Increasing the latter may require better compressor materials.

Overall pressure ratio can be increased by improving fan (or) LP compressor pressure ratio and/or HP compressor pressure ratio. If the latter is held constant, the increase in (HP) compressor delivery temperature (from raising overall pressure ratio) implies an increase in HP mechanical speed. However, stressing considerations might limit this parameter, implying, despite an increase in overall pressure ratio, a reduction in HP compressor pressure ratio.

According to simple theory, if the ratio of turbine rotor inlet temperature/(HP) compressor delivery temperature is maintained, the HP turbine throat area can be retained. However, this assumes that cycle improvements are obtained, while retaining the datum (HP) compressor exit flow function (non-dimensional flow). In practice, changes to the non-dimensional speed of the (HP) compressor and cooling bleed extraction would probably make this assumption invalid, making some adjustment to HP turbine throat area unavoidable. This means the HP turbine nozzle guide vanes would have to be different from the original. In all probability, the downstream LP turbine nozzle guide vanes would have to be changed anyway.

Thrust Growth

Thrust growth is obtained by increasing core power. There are two basic routes available:

- hot route: increase HP turbine rotor inlet temperature
- cold route: increase core mass flow

Both routes require an increase in the combustor fuel flow and, therefore, the heat energy added to the core stream.

The hot route may require changes in turbine blade/vane materials and/or better blade/vane cooling. The cold route can be obtained by one of the following:

- adding T-stages to the LP/IP compression

- adding a zero-stage to the HP compression

- improving the compression process, without adding stages (e.g. higher fan hub pressure ratio)

all of which increase both overall pressure ratio and core airflow.

Alternatively, the core size can be increased, to raise core airflow, without changing overall pressure ratio. This route is expensive, since a new (upflowed) turbine system (and possibly a larger IP compressor) is also required.

Changes must also be made to the fan to absorb the extra core power. On a civil engine, jet noise considerations mean that any significant increase in take-off thrust must be accompanied by a corresponding increase in fan mass flow (to maintain a T/O specific thrust of about 30 lbf/ lb/s). To reduce noise civilian turbofans have a specially shaped nozzle that limits the exhaust speed to subsonic speeds. This leads to a thermic clogging termed *choked nozzle* where the mass flow cannot be increased beyond a certain amount. Thus, the mass flow can only be increased through the bypass airstream, usually by increasing fan diameter. On military engines, the fan pressure ratio would probably be increased to improve specific thrust, jet noise not normally being an important factor.

Technical Discussion

1. Specific thrust (net thrust/intake airflow) is an important parameter for turbofans and jet engines in general. Imagine a fan (driven by an appropriately sized electric motor) operating within a pipe, which is connected to a propelling nozzle. It is fairly obvious, the higher the fan pressure ratio (fan discharge pressure/fan inlet pressure), the higher the jet velocity and the corresponding specific thrust. Now imagine we replace this set-up with an equivalent turbofan - same airflow and same fan pressure ratio. Obviously, the core of the turbofan must produce sufficient power to drive the fan via the low pressure (LP) turbine. If we choose a low (HP) turbine inlet temperature for the gas generator, the core airflow needs to be relatively high to compensate. The corresponding bypass ratio is therefore relatively low. If we raise the turbine inlet temperature, the core airflow can be smaller, thus increasing bypass ratio. Raising turbine inlet temperature tends to increase thermal efficiency and, therefore, improve fuel efficiency.

2. Naturally, as altitude increases, there is a decrease in air density and, therefore, the net thrust of an engine. There is also a flight speed effect, termed thrust lapse rate. Consider the approximate equation for net thrust again:

$$F_n = m \cdot (V_{jfe} - V_a)$$

3. With a high specific thrust (e.g., fighter) engine, the jet velocity is relatively high, so intuitively one can see that increases in flight velocity have less of an impact upon net thrust than a medium specific thrust (e.g., trainer) engine, where the jet velocity is lower. The impact of thrust lapse rate upon a low specific thrust (e.g., civil) engine is even more severe. At high flight speeds, high-specific-thrust engines can pick up net thrust through the ram

rise in the intake, but this effect tends to diminish at supersonic speeds because of shock wave losses.

4. Thrust growth on civil turbofans is usually obtained by increasing fan airflow, thus preventing the jet noise becoming too high. However, the larger fan airflow requires more power from the core. This can be achieved by raising the overall pressure ratio (combustor inlet pressure/intake delivery pressure) to induce more airflow into the core and by increasing turbine inlet temperature. Together, these parameters tend to increase core thermal efficiency and improve fuel efficiency.

5. Some high bypass ratio civil turbofans use an extremely low area ratio (less than 1.01), convergent-divergent, nozzle on the bypass (or mixed exhaust) stream, to control the fan working line. The nozzle acts as if it has variable geometry. At low flight speeds the nozzle is unchoked (less than a Mach number of unity), so the exhaust gas speeds up as it approaches the throat and then slows down slightly as it reaches the divergent section. Consequently, the nozzle exit area controls the fan match and, being larger than the throat, pulls the fan working line slightly away from surge. At higher flight speeds, the ram rise in the intake increases nozzle pressure ratio to the point where the throat becomes choked (M=1.0). Under these circumstances, the throat area dictates the fan match and, being smaller than the exit, pushes the fan working line slightly towards surge. This is not a problem, since fan surge margin is much better at high flight speeds.

6. The off-design behaviour of turbofans is illustrated under compressor map and turbine map.

7. Because modern civil turbofans operate at low specific thrust, they only require a single fan stage to develop the required fan pressure ratio. The desired overall pressure ratio for the engine cycle is usually achieved by multiple axial stages on the core compression. Rolls-Royce tend to split the core compression into two with an intermediate pressure (IP) supercharging the HP compressor, both units being driven by turbines with a single stage, mounted on separate shafts. Consequently, the HP compressor need only develop a modest pressure ratio (e.g., ~4.5:1). US civil engines use much higher HP compressor pressure ratios (e.g., ~23:1 on the General Electric GE90) and tend to be driven by a two-stage HP turbine. Even so, there are usually a few IP axial stages mounted on the LP shaft, behind the fan, to further supercharge the core compression system. Civil engines have multi-stage LP turbines, the number of stages being determined by the bypass ratio, the amount of IP compression on the LP shaft and the LP turbine blade speed.

8. Because military engines usually have to be able to fly very fast at sea level, the limit on HP compressor delivery temperature is reached at a fairly modest design overall pressure ratio, compared with that of a civil engine. Also the fan pressure ratio is relatively high, to achieve a medium to high specific thrust. Consequently, modern military turbofans usually only have 5 or 6 HP compressor stages and only require a single-stage HP turbine. Low bypass ratio military turbofans usually have one LP turbine stage, but higher bypass ratio engines need two stages. In theory, by adding IP compressor stages, a modern military turbofan HP compressor could be used in a civil turbofan derivative, but the core would tend to be too small for high thrust applications.

Engine Noise

Turbofan engine noise propagates both upstream the inlet and downstream the primary nozzle and the by-pass duct. The main noise sources are the turbine and the compressor, the jet and the fan. The contribution of each noise source significantly evolved in the last decades: in typical 1960s design the jet was the main source whereas in modern turbofans the fan is the main noise source.

The fan noise is a tonal noise and its signature depends on the fan rotational speed:

at low speed, the fan noise is due to the interaction of the blades with the distorted flow injected in the engine; this happens for example during the approach;

at high engine ratings, the fan tip is supersonic and this allows intense rotor-locked duct modes to propagate upstream; this noise is known as "buzz saw" and is typical at take-off.

All modern turbofan engines are equipped with acoustic liners to damp the noise generated. These are installed in the nacelle, and they extend as much as possible to cover the largest area. The acoustic performance of the engine can be experimentally evaluated by means of ground tests or in dedicated experimental test rigs.

Recent Developments in Blade Technology

The turbine blades in a turbofan engine are subject to high heat and stress, and require special fabrication. New material construction methods and material science have allowed blades, which were originally polycrystalline (regular metal), to be made from lined up metallic crystals and more recently mono-crystalline (i.e., single crystal) blades, which can operate at higher temperatures with less distortion.

Nickel-based superalloys are used for HP turbine blades in almost all modern jet engines. The temperature capabilities of turbine blades have increased mainly through four approaches: the manufacturing (casting) process, cooling path design, thermal barrier coating (TBC), and alloy development.

Although turbine blade (and vane) materials have improved over the years, much of the increase in (HP) turbine inlet temperatures is due to improvements in blade/vane cooling technology. Relatively cool air is bled from the compression system, bypassing the combustion process, and enters the hollow blade or vane. The gas temperature can therefore be even higher than the melting temperature of the blade. After picking up heat from the blade/vane, the cooling air is dumped into the main gas stream. If the local gas temperatures are low enough, downstream blades/vanes are uncooled and not adversely affected.

Strictly speaking, cycle-wise the HP turbine rotor inlet temperature (after the temperature drop across the HPT stator) is more important than the (HP) turbine inlet temperature. Although some modern military and civil engines have peak RITs of the order of 1,560 °C (2,840 °F), such temperatures are only experienced for a short time (during take-off) on civil engines.

Turbofan Engine Manufacturers

The turbofan engine market is dominated by General Electric, Rolls-Royce plc and Pratt & Whitney, in order of market share. GE and SNECMA of France have a joint venture, CFM International. Pratt & Whitney also have a joint venture, International Aero Engines with Japanese Aero Engine

Corporation and MTU of Germany, specializing in engines for the Airbus A320 family. Pratt & Whitney and General Electric have a joint venture, Engine Alliance selling a range of engines for aircraft such as the Airbus A380.

General Electric

GE Aviation, part of the General Electric Conglomerate, currently has the largest share of the turbofan engine market. Some of their engine models include the CF6 (available on the Boeing 767, Boeing 747, Airbus A330 and more), GE90 (only the Boeing 777) and GEnx (developed for the Boeing 747-8 & Boeing 787 Dreamliner and proposed for the Airbus A350, currently in development) engines. On the military side, GE engines power many U.S. military aircraft, including the F110, powering 80% of the US Air Force's F-16 Fighting Falcons, and the F404 and F414 engines, which power the Navy's F/A-18 Hornet and Super Hornet. Rolls-Royce and General Electric were jointly developing the F136 engine to power the Joint Strike Fighter, however, due to government budget cuts, the program has been eliminated.

Rolls-Royce

Rolls-Royce plc is the second largest manufacturer of turbofans and is most noted for their RB211 and Trent series, as well as their joint venture engines for the Airbus A320 and McDonnell Douglas MD-90 families (IAE V2500 with Pratt & Whitney and others), the Panavia Tornado (Turbo-Union RB199) and the Boeing 717 (BR700). The Rolls-Royce AE 3007, developed by Allison Engine Company before its acquisition by Rolls-Royce, powers several Embraer regional jets. Rolls-Royce Trent 970s were the first engines to power the new Airbus A380. The famous thrust vectoring Pegasus - actually a Bristol Siddeley design taken on by Rolls-Royce when they took over that company - is the primary powerplant of the Harrier "Jump Jet" and its derivatives.

Pratt & Whitney

Pratt & Whitney is third behind GE and Rolls-Royce in market share. The JT9D has the distinction of being chosen by Boeing to power the original Boeing 747 "Jumbo jet". The PW4000 series is the successor to the JT9D, and powers some Airbus A310, Airbus A300, Boeing 747, Boeing 767, Boeing 777, Airbus A330 and MD-11 aircraft. The PW4000 is certified for 180-minute ETOPS when used in twinjets. The first family has a 94-inch (2.4 m) fan diameter and is designed to power the Boeing 767, Boeing 747, MD-11, and the Airbus A300. The second family is the 100 inch (2.5 m) fan engine developed specifically for the Airbus A330 twinjet, and the third family has a diameter of 112-inch (2.8 m) designed to power Boeing 777. The Pratt & Whitney F119 and its derivative, the F135, power the United States Air Force's F-22 Raptor and the international F-35 Lightning II, respectively. Rolls-Royce are responsible for the lift fan which will provide the F-35B variants with a STOVL capability. The F100 engine was first used on the F-15 Eagle and F-16 Fighting Falcon. Newer Eagles and Falcons also come with GE F110 as an option, and the two are in competition.

CFM International

CFM International is a joint venture between GE Aircraft Engines and SNECMA of France. They have created the very successful CFM56 series, used on Boeing 737, Airbus A340, and Airbus A320 family aircraft.

Engine Alliance

Engine Alliance is a 50/50 joint venture between General Electric and Pratt & Whitney formed in August 1996 to develop, manufacture, sell, and support a family of modern technology aircraft engines for new high-capacity, long-range aircraft. The main application for such an engine, the GP7200, was originally the Boeing 747-500/600X projects, before these were cancelled owing to lack of demand from airlines. Instead, the GP7000 has been re-optimised for use on the Airbus A380 superjumbo. In that market it is competing with the Rolls-Royce Trent 900, the launch engine for the aircraft. The two variants are the GP7270 and the GP7277.

International Aero Engines

International Aero Engines is a Zürich-registered joint venture between Pratt & Whitney, MTU Aero Engines and Japanese Aero Engine Corporation. The collaboration produced the V2500, the second most successful commercial jet engine program in production today in terms of volume, and the third most successful commercial jet engine program in aviation history.

Williams International

Williams International is a manufacturer of small gas turbine engines based in Walled Lake, Michigan, United States. It produces jet engines for cruise missiles and small jet-powered aircraft. They have been producing engines since the 1970s and the range produces between 1000 and 3600 pounds of thrust. The engines are used as original equipment on the Cessna CitationJet CJ1 through CJ4 and Cessna Mustang, Beechcraft 400XPR and Premier 1a and there are several development programs with other manufacturers. The range is also very popular with the re-engine market being used by Sierra Jet and Nextant to breathe new life into aging platforms.

Honeywell Aerospace

Honeywell Aerospace is one of the largest manufacturer of aircraft engines and avionics, as well as a producer of auxiliary power units (APUs) and other aviation products. Headquartered in Phoenix, Arizona, it is a division of the Honeywell International conglomerate. Honeywell/ITEC F124 series is used in military jets, such as the Aero L-159 Alca and the Alenia Aermacchi M-346. The Honeywell HTF700 series is used in the Bombardier Challenger 300 and the Gulfstream G280. The ALF502 and LF507 turbofans are produced by a partnership between Honeywell and China's state-owned Industrial Development Corporation. The partnership is called the International Turbine Engine Co.

Aviadvigatel

Aviadvigatel is a Russian manufacturer of aircraft engines that succeeded the Soviet Soloviev Design Bureau. The company currently offers several versions of the Aviadvigatel PS-90 engine that powers Ilyushin Il-96-300/400/400T, Tupolev Tu-204, Tu-214 series and the Ilyushin Il-76-MD-90. The company is also developing the new Aviadvigatel PD-14 engine for the new Russian MS-21 airliner.

Ivchenko-Progress

Ivchenko-Progress is the Ukrainian aircraft engine company that succeeded the Soviet Ivchenko

Design Bureau. Some of their engine models include Progress D-436 available on the Antonov An-72/74, Yakovlev Yak-42, Beriev Be-200, Antonov An-148 and Tupolev Tu-334 and Progress D-18T that powers two of the world's largest airplanes, Antonov An-124 and Antonov An-225.

NPO Saturn

NPO Saturn is a Russian aircraft engine manufacturer, formed from the mergers of Rybinsk and Lyul'ka-Saturn. Saturn's engines include Lyulka AL-31, Lyulka AL-41, NPO Saturn AL-55 and power many former Eastern Bloc aircraft, such as the Tupolev Tu-154. Saturn holds a 50% stake in the PowerJet joint venture with Snecma.

PowerJet

PowerJet is a 50-50 joint venture between Snecma (Safran) and NPO Saturn, created in July 2004. The company manufactures SaM146, the sole powerplant for the Sukhoi Superjet 100.

Klimov

Klimov was formed in the early 1930s to produce and improve upon the liquid-cooled Hispano-Suiza 12Y V-12 piston engine for which the USSR had acquired a license. Currently, Klimov is the manufacturer of the Klimov RD-33 turbofan engines.

EuroJet

EuroJet Turbo GmbH is a multi-national consortium, the partner companies of which are Rolls Royce of the United Kingdom, Avio of Italy, ITP of Spain and MTU Aero Engines of Germany. Eurojet GmbH was formed in 1986 to manage the development, production, support, maintenance, support and sales of the EJ200 turbofan engine for the Eurofighter Typhoon.

Chinese Turbofans

Three Chinese corporations build turbofan engines. Some of these are licensed or reverse engineered versions of European and Russian turbofans, and the other are indigenous models, but all are in development phase. Shenyang Aircraft Corporation (manufacturer of Shenyang WS-10), Xi'an Aero-Engine Corporation (manufacturer of Xian WS-15) and Guizhou Aircraft Industry Corporation (manufacturer of Guizhou WS-13) manufacture turbofans.

Japanese Turbofans

Ishikawajima-Harima Heavy Industries is the Japan aircraft engine company. The company manufactures F3 for Kawasaki T-4, XF5-1 for ATD-X, F7 for Kawasaki P-1.

Gas Turbine Research Establishment (GTRE)

Gas Turbine Research Establishment is owned by DRDO of Government of India. It produced the GTRE GTX-35VS Kaveri turbofan intended to power HAL Tejas and HAL Advanced Medium Combat Aircraft being built by the Aeronautical Development Agency.

General Electric CF6 which powers the Airbus A300, Boeing 747, Douglas DC-10 and other aircraft

Rolls-Royce Trent 900 undergoing climatic testing

Pratt & Whitney PW4000 which powered the first Boeing 777

Commercial turbofans in production							
Model	Start	Bypass	Length	Fan	Weight	Thrust	Major applications
GE GE90	1992	8.7–9.9	5.18m–5.40m	3.12–3.25 m	7.56–8.62t	330–510 kN	B777
P&W PW4000	1984	4.8–6.4	3.37–4.95m	2.84 m	4.18–7.48t	222–436 kN	A300/A310, A330, B747, B767, B777, MD-11
R-R Trent XWB	2010	9.3	5.22 m	3.00 m	7.28 t	330–430 kN	A350XWB
R-R Trent 800	1993	5.7–5.79	4.37m	2.79m	5.96–5.98t	411–425 kN	B777
EA GP7000	2004	8.7	4.75 m	2.95 m	6.09-6.71 t	311-363 kN	A380
R-R Trent 900	2004	8.7	4.55 m	2.95 m	6.18-6.25 t	340-357 kN	A380
R-R Trent 1000	2006	10.8–11	4.74 m	2.85 m	5.77 t	240–350 kN	B787
GE GEnx	2006	8.0-9.3	4.31-4.69 m	2.66-2.82 m	5.62-5.82 t	296-339 kN	B747-8, B787
R-R Trent 700	1990	4.9	3.91 m	2.47 m	4.79 t	320 kN	A330
GE CF6	1971	4.3-5.3	4.00-4.41 m	2.20-2.79 m	3.82-5.08 t	222-298 kN	A300/A310, A330, B747, B767, MD-11
R-R Trent 500	1999	8.5	3.91 m	2.47 m	4.72 t	252 kN	A340
P&W PW1000G	2008	9.0-12.5	3.40 m	1.42-2.06 m	2.86 t	67-160 kN	A320neo, CSeries, E-Jets E2
CFM LEAP	2013	9.0-11.0	3.15-3.33m	1.76-1.98m	2.78-3.15t	100-146 kN	A320neo, B737Max
CFM56	1974	5.0-6.6	2.36-2.52m	1.52-1.84m	1.95-2.64t	97.9-151 kN	A320, A340, B737, KC-135
IAE V2500	1987	4.4-4.9	3.20m	1.60m	2.36-2.54t	97.9-147 kN	A320, MD-90
P&W PW6000	2000	4.90	2.73m	1.44m	2.36t	100.2 kN	Airbus A318
R-R BR700	1994	4.2-4.5	3.41-3.60m	1.32-1.58m	1.63-2.11t	68.9-102.3 kN	B717, Global Express, Gulfstream V
GE Passport	2013	5.6	3.37m	1.30m	2.07t	78.9–84.2 kN	Global 7000/8000
GE CF34	1982	5.3-6.3	2.62-3.26m	1.25-1.32m	0.74-1.12t	41-82.3 kN	Challenger 600, CRJ, E-jets
P&WC PW800	2012	5.5		1.30m		67.4-69.7 kN	Gulfstream G500/G600
R-R Tay	1984	3.1-3.2	2.41m	1.12-1.14m	1.42-1.53t	61.6-68.5 kN	Gulfstream IV, Fokker 70/100
Silvercrest	2012	5.9	1.90m	1.08m	1.09t	50.9 kN	Cit. Hemisphere, Falcon 5X
R-R AE 3007	1991	5.0	2.71m	1.11m	0.72t	33,7 kN	ERJ, Citation X
P&WC PW300	1988	3.8-4.5	1.92-2.07	0.97m	0.45-0.47t	23.4-35.6 kN	Cit. Sovereign, G200, F. 7X, F. 2000

Commercial turbofans in production							
Model	Start	Bypass	Length	Fan	Weight	Thrust	Major applications
HW HTF7000	1999	4.4	2.29m	0.87m	0.62t	28.9 kN	Challenger 300, G280, Legacy 500
HW TFE731	1970	2.66-3.9	1.52-2.08m	.072-0.78m	0.34-0.45t	15.6-22.2 kN	Learjet 70/75, G150, Falcon 900
Williams FJ44	1985	3.3-4.1	1.36-2.09m	.53-0.57m	0.21-0.24t	6.7-15.6 kN	CitationJet, Cit. M2
P&WC PW500	1993	3.90	1.52m	0.70m	0.28t	13.3 kN	Citation Excel, Phenom 300
GE-H HF120	2009	4.43	1.12m	0.54 m	0.18t	7.4 kN	HondaJet
Williams FJ33	1998		0.98m	0.53 m	0.14 t	6.7 kN	Cirrus SF50
P&WC PW600	2001	1.8-2.8	0.67m	0.36m	0.15t	6.0 kN	Cit. Mustang, Eclipse 500, Phenom 100

Extreme Bypass Jet Engines

In the 1970s, Rolls-Royce/SNECMA tested a M45SD-02 turbofan fitted with variable pitch fan blades to improve handling at ultra low fan pressure ratios and to provide thrust reverse down to zero aircraft speed. The engine was aimed at ultra quiet STOL aircraft operating from city centre airports.

In a bid for increased efficiency with speed, a development of the *turbofan* and *turboprop* known as a propfan engine was created that had an unducted fan. The fan blades are situated outside of the duct, so that it appears like a turboprop with wide scimitar-like blades. Both General Electric and Pratt & Whitney/Allison demonstrated propfan engines in the 1980s. Excessive cabin noise and relatively cheap jet fuel prevented the engines being put into service.

Terminology

Afterburner

> extra combustor immediately upstream of final nozzle (also called reheat)

Augmentor

> afterburner on low-bypass turbofan engines.

Average stage loading

> constant × (delta temperature)/[(blade speed) × (blade speed) × (number of stages)]

Bypass

> airstream that completely bypasses the core compression system, combustor and turbine system

Bypass ratio

bypass airflow /core compression inlet airflow

Core

turbomachinery handling the airstream that passes through the combustor.

Core power

residual shaft power from ideal turbine expansion to ambient pressure after deducting core compression power

Core thermal efficiency

core power/power equivalent of fuel flow

Dry

afterburner (if fitted) not lit

EGT

exhaust gas temperature

EPR

engine pressure ratio

Fan

turbofan LP compressor

Fan pressure ratio

fan outlet total pressure/intake delivery total pressure

Flex temp

use of artificially high apparent air temperature to reduce engine wear

Gas generator

engine core

HP compressor

high-pressure compressor (also HPC)

HP turbine

high-pressure turbine

Intake ram drag

penalty associated with jet engines picking up air from the atmosphere (conventional rocket motors do not have this drag term, because the oxidiser travels with the vehicle)

IEPR

integrated engine pressure ratio

IP compressor

intermediate pressure compressor (also IPC)

IP turbine

intermediate pressure turbine (also IPT)

LP compressor

low-pressure compressor (also LPC)

LP turbine

low-pressure turbine (also LPT)

Net thrust

nozzle total gross thrust - intake ram drag (excluding nacelle drag, etc., this is the basic thrust acting on the airframe)

Overall pressure ratio

combustor inlet total pressure/intake delivery total pressure

Overall efficiency

thermal efficiency * propulsive efficiency

Propulsive efficiency

propulsive power/rate of production of propulsive kinetic energy (maximum propulsive efficiency occurs when jet velocity equals flight velocity, which implies zero net thrust!)

Specific fuel consumption (SFC)

total fuel flow/net thrust (proportional to flight velocity/overall thermal efficiency)

Spooling up

accelerating, marked by a delay

Static pressure

pressure of the fluid which is associated not with its motion but with its state

Specific thrust

net thrust/intake airflow

Thermal efficiency

rate of production of propulsive kinetic energy/fuel power

Total fuel flow

combustor (plus any afterburner) fuel flow rate (e.g., lb/s or g/s)

Total pressure

static pressure plus kinetic energy term

Turbine rotor inlet temperature

gas absolute mean temperature at principal (e.g., HP) turbine rotor entry

Turbopump

This article is about the fuel pump. For turbine devices for producing a high vacuum, see Turbomolecular pump.

A turbopump is a propellant pump with two main components: a rotodynamic pump and a driving gas turbine, usually both mounted on the same shaft, or sometimes geared together. The purpose of a turbopump is to produce a high-pressure fluid for feeding a combustion chamber or other use.

An axial turbopump designed and built for the M-1 rocket engine

A turbopump can comprise one of two types of pumps: a centrifugal pump, where the pumping is done by throwing fluid outward at high speed, or an axial-flow pump, where alternating rotating and static blades progressively raise the pressure of a fluid.

Axial-flow pumps have small diameters but give relatively modest pressure increases. Although multiple compression stages are needed, axial flow pumps work well with low-density fluids. Centrifugal pumps are far more powerful for high-density fluids but require large diameters for low-density fluids.

Turbopumps operate in much the same way as turbocharger units for vehicles: higher fuel pressures allow fuel to be supplied to higher-pressure combustion chambers for higher-performance engines.

History

The V-2 rocket used a circular turbopump to pressurize the propellant.

Early Development

High-pressure pumps for larger missiles had been discussed by rocket pioneers such as Hermann Oberth. In mid-1935 Wernher von Braun initiated a fuel pump project at the southwest German firm *Klein, Schanzlin & Becker* that was experienced in building large fire-fighting pumps. The V-2 rocket design used hydrogen peroxide decomposed through a Walter steam generator to power the uncontrolled turbopump produced at the Heinkel plant at Jenbach, so V-2 turbopumps and combustion chamber were tested and matched to prevent the pump from overpressurizing the chamber. The first engine fired successfully in September, and on August 16, 1942, a trial rocket stopped in mid-air and crashed due to a failure in the turbopump. The first successful V-2 launch was on October 3, 1942.

Development from 1947 to 1949

The principal engineer for turbopump development at Aerojet was George Bosco. During the second half of 1947, Bosco and his group learned about the pump work of others and made preliminary design studies. Aerojet representatives visited Ohio State University where Florant was working on hydrogen pumps, and consulted Dietrich Singelmann, a German pump expert at Wright Field. Bosco subsequently used Singelmann's data in designing Aerojet's first hydrogen pump.

By mid-1948, Aerojet had selected centrifugal pumps for both liquid hydrogen and liquid oxygen. They obtained some German radial-vane pumps from the Navy and tested them during the second half of the year.

By the end of 1948, Aerojet had designed, built, and tested a liquid hydrogen pump (15 cm diameter). Initially, it used ball bearings that were run clean and dry, because the low temperature made conventional lubrication impractical. The pump was first operated at low speeds to allow its parts to cool down to operating temperature. When temperature gauges showed that liquid hydrogen had reached the pump, an attempt was made to accelerate from 5000 to 35 000 revolutions per minute. The pump failed and examination of the pieces pointed to a failure of the bearing, as well as the impeller. After some testing, super-precision bearings, lubricated by oil that was atomized and directed by a stream of gaseous nitrogen, were used. On the next run, the bearings worked satisfactorily but the stresses were too great for the brazed impeller and it flew apart. A new one was made by milling from a solid block of aluminum. The next two runs with the new pump were a great disappointment; the instruments showed no significant flow or pressure rise. The problem was traced to the exit diffuser of the pump, which was too small and insufficiently cooled during the cool-down cycle so that it limited the flow. This was corrected by adding vent holes in the pump housing; the vents were opened during cool down and closed when the pump was cold. With this fix, two additional runs were made in March 1949 and both were successful. Flow rate and pressure were found to be in approximate agreement with theoretical predictions. The maximum pressure was 26 atmospheres and the flow was 0.25 kilogram per second.

The Space Shuttle Main Engine's turbopumps spun at over 30,000 rpm, delivering 150 lb (68 kg) of liquid hydrogen and 896 lb (406 kg) of liquid oxygen to the engine per second.

Centrifugal Turbopumps

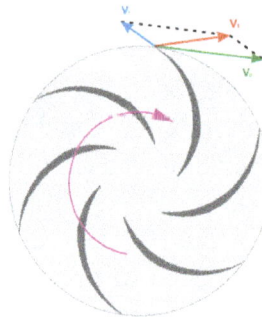

In centrifugal turbopumps a rotating disk throws the fluid to the rim

Most turbopumps are centrifugal - the fluid enters the pump near the axis and the rotor accelerates the fluid to high speed. The fluid then passes through a diffuser which is a progressively enlarging pipe, which permits recovery of the dynamic pressure. The diffuser turns the high kinetic energy into high pressures (hundreds of bar is not uncommon), and if the outlet backpressure is not too high, high flow rates can be achieved.

Axial Turbopumps

Axial compressors

Axial turbopumps also exist. In this case the axle essentially has propellers attached to the shaft, and the fluid is forced by these parallel with the main axis of the pump. Generally, axial pumps tend to give much lower pressures than centrifugal pumps, and a few bars is not uncommon. They are, however, still useful – axial pumps are commonly used as "inducers" for centrifugal pumps, which raise the inlet pressure of the centrifugal pump enough to prevent excessive cavitation from occurring therein.

Complexities of Centrifugal Turbopumps

Turbopumps have a reputation for being extremely hard to design to get optimal performance. Whereas a well engineered and debugged pump can manage 70–90% efficiency, figures less than half that are not uncommon. Low efficiency may be acceptable in some applications, but in rocketry this is a severe problem. Turbopumps in rockets are important and problematic enough that launch vehicles using one have been caustically described as a "turbopump with a rocket attached"–up to 55% of the total cost has been ascribed to this area.

Common problems include:

- excessive flow from the high-pressure rim back to the low-pressure inlet along the gap between the casing of the pump and the rotor,

- excessive recirculation of the fluid at inlet,

- excessive vortexing of the fluid as it leaves the casing of the pump,

- damaging cavitation to impeller blade surfaces in low-pressure zones.

In addition, the precise shape of the rotor itself is critical.

Driving Turbopumps

Steam turbine-powered turbopumps are employed when there is a source of steam, e.g. the boilers of steam ships. Gas turbines are usually used when electricity or steam is not available and place or weight restrictions permit the use of more efficient sources of mechanical energy.

One of such cases are rocket engines, which need to pump fuel and oxidizer into their combustion chamber. This is necessary for large liquid rockets, since forcing the fluids or gases to flow by simple pressurizing of the tanks is often not feasible; the high pressure needed for the required flow rates would need strong and heavy tanks.

Ramjet motors are also usually fitted with turbopumps, the turbine being driven either directly by external freestream ram air or internally by airflow diverted from combustor entry. In both cases the turbine exhaust stream is dumped overboard.

Turboprop

A turboprop engine is a turbine engine that drives an aircraft propeller. In contrast to a turbojet, the engine's exhaust gases do not contain enough energy to create significant thrust, since almost all of the engine's power is used to drive the propeller.

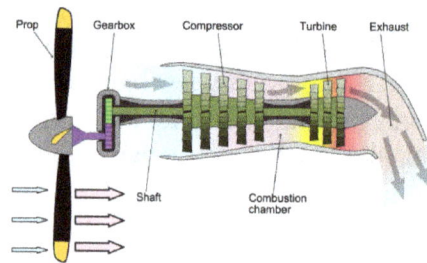

Schematic diagram showing the operation of a turboprop engine

A Fairchild F-27 representative of the 2nd generation of modern Rolls-Royce Dart turboprop powered propjet aircraft, after the initial success of the 1950s era Vickers Viscount.

An ATR-72, a typical modern turboprop aircraft.

The propeller is coupled to the turbine through a reduction gear that converts the high RPM, low torque output to low RPM, high torque. The propeller itself is normally a constant speed (variable pitch) type similar to that used with larger reciprocating aircraft engines.

Turboprop engines are generally used on small subsonic aircraft, but some aircraft outfitted with turboprops have cruising speeds in excess of 500 kt (926 km/h, 575 mph). Large military and civil aircraft, such as the Lockheed L-188 Electra and the Tupolev Tu-95, have also used turboprop power. The Airbus A400M is powered by four Europrop TP400 engines, which are the third most powerful turboprop engines ever produced, after the eleven megawatt-output Kuznetsov NK-12 and 10.4 MW-output Progress D-27.

In its simplest form a turboprop consists of an intake, compressor, combustor, turbine, and a propelling nozzle. Air is drawn into the intake and compressed by the compressor. Fuel is then added to the compressed air in the combustor, where the fuel-air mixture then combusts. The hot combustion gases expand through the turbine. Some of the power generated by the turbine is used to drive the compressor. The rest is transmitted through the reduction gearing to the propeller. Further expansion of the gases occurs in the propelling nozzle, where the gases exhaust to atmospheric pressure. The propelling nozzle provides a relatively small proportion of the thrust generated by a turboprop.

Turboprops are most efficient at flight speeds below 725 km/h (450 mph; 390 knots) because the jet velocity of the propeller (and exhaust) is relatively low. Due to the high price of turboprop engines, they are mostly used where high-performance short-takeoff and landing (STOL) capability and efficiency at modest flight speeds are required. The most common application of turboprop engines in civilian aviation is in small commuter aircraft, where their greater power and reliability over reciprocating engines offsets their higher initial cost and fuel consumption. Turboprop airliners now operate at near the same speed as small turbofan-powered aircraft but burn two-thirds of the fuel per passenger. However, compared to a turbojet (which can fly at high altitude for enhanced speed and fuel efficiency) a propeller aircraft has a much lower ceiling. Turboprop-powered aircraft have become popular for bush airplanes such as the Cessna Caravan and Quest Kodiak as jet fuel is easier to obtain in remote areas than is aviation-grade gasoline (avgas).

Technological Aspects

Exhaust thrust in a turboprop is sacrificed in favor of shaft power, which is obtained by extracting additional power (up to that necessary to drive the compressor) from turbine expansion. Owing to the additional expansion in the turbine system, the residual energy in the exhaust jet is low. Consequently, the exhaust jet produces (typically) less than 10% of the total thrust, and turboprops can have bypass ratios up to 50-100 although the propulsion airflow is less clearly defined for propellers than for fans.

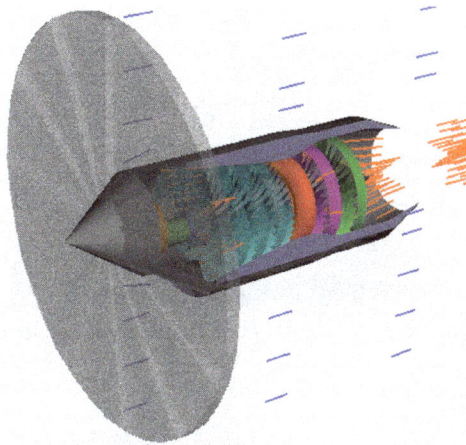

Flow past a turboprop engine in operation

Unlike the small diameter fans used in turbofan jet engines, the propeller has a large diameter that lets it accelerate a large volume of air. This permits a lower airstream velocity for a given amount of thrust. As it is more efficient at low speeds to accelerate a large amount of air by a small degree than a small amount of air by a large degree, a low disc loading (thrust per disc area) increases the aircraft's energy efficiency, and this reduces the fuel use.

Propellers lose efficiency as aircraft speed increases, so turboprops are normally not used on high-speed aircraft above Mach 0.6-0.7. However, propfan engines, which are very similar to turboprop engines, can cruise at flight speeds approaching Mach 0.75. To increase propeller efficiency, a mechanism can be used to alter their pitch relative to the airspeed. A variable-pitch propeller, also called a controllable-pitch propeller, can also be used to generate negative thrust while decelerating on the runway. Additionally, in the event of an engine outage, the pitch can be adjusted to a vaning pitch (called feathering), thus minimizing the drag of the non-functioning propeller.

While most modern turbojet and turbofan engines use axial-flow compressors, turboprop engines usually contain at least one stage of centrifugal compression. Centrifugal compressors have the advantage of being simple and lightweight, at the expense of a streamlined shape.

While the power turbine may be integral with the gas generator section, many turboprops today feature a free power turbine on a separate coaxial shaft. This enables the propeller to rotate freely, independent of compressor speed. Residual thrust on a turboshaft is avoided by further expansion in the turbine system and/or truncating and turning the exhaust 180 degrees, to produce two opposing jets. Apart from the above, there is very little difference between a turboprop and a turboshaft.

Some commercial aircraft with turboprop engines include the Bombardier Dash 8, ATR 42, ATR 72, BAe Jetstream 31, Beechcraft 1900, Embraer EMB 120 Brasilia, Fairchild Swearingen Metroliner, Dornier 328, Saab 340 and 2000, Xian MA60, Xian MA600, and Xian MA700, Fokker 27, 50 and 60.

History

A Rolls-Royce RB.50 *Trent* on a test rig at Hucknall, in March 1945

Kuznetsov NK-12M Turboprop, on a Tu-95

Alan Arnold Griffith had published a paper on turbine design in 1926. Subsequent work at the Royal Aircraft Establishment investigated axial turbine designs that could be used to supply power to a shaft and thence a propeller. From 1929, Frank Whittle began work on centrifugal turbine designs that would deliver pure jet thrust.

The world's first turboprop was designed by the Hungarian mechanical engineer György Jendrassik. Jendrassik published a turboprop idea in 1928, and on 12 March 1929 he patented his invention. In 1938, he built a small-scale (100 Hp; 74.6 kW) experimental gas turbine. The larger Jendrassik Cs-1, with a predicted output of 1,000 bhp, was produced and tested at the Ganz Works in Budapest between 1937 and 1941. It was of axial-flow design with 15 compressor and 7 turbine stages, annular combustion chamber and many other modern features. First run in 1940, combustion problems limited its output to 400 bhp. In 1941,the engine was abandoned due to war, and the factory was turned over to conventional engine production. The world's first turboprop engine that went into mass production was designed by a German engineer, Max Adolf Mueller, in 1942.

The first public mention of turboprop engine in a general public press, was in the British aviation publication, *Flight*, in February 1944 issue, which included a detailed cutaway drawing of what a possible future turboprop engine could look like. The drawing was very close to what the future Rolls-Royce Trent would look like. The first British turboprop engine was the Rolls-Royce RB.50 Trent, a converted Derwent II fitted with reduction gear and a Rotol 7-ft, 11-in five-bladed propeller. Two Trents were fitted to Gloster Meteor *EE227* — the sole "Trent-Meteor" — which thus became the world's first turboprop-powered aircraft, albeit a test-bed not intended for production. It first flew on 20 September 1945. From their experience with the Trent, Rolls-Royce developed the Rolls-Royce Clyde, the first turboprop engine to be fully type certificated for military and civil use, and the Dart, which became one of the most reliable turboprop engines ever built. Dart production continued for more than fifty years. The Dart-powered Vickers Viscount was the first turboprop aircraft of any kind to go into production and sold in large numbers. It was also the first four-engined turboprop. Its first flight was on 16 July 1948. The world's first single engined turboprop aircraft was the Armstrong Siddeley Mamba-powered Boulton Paul Balliol, which first flew on 24 March 1948.

Rolls-Royce Dart turboprop engine

The Soviet Union built on German World War II development by Junkers Motorenwerke, while BMW, Heinkel-Hirth and Daimler-Benz also developed and partially tested designs. While the

Soviet Union had the technology to create the airframe for a jet-powered strategic bomber comparable to Boeing's B-52 Stratofortress, they instead produced the Tupolev Tu-95 Bear, powered with four Kuznetsov NK-12 turboprops, mated to eight contra-rotating propellers (two per nacelle) with supersonic tip speeds to achieve maximum cruise speeds in excess of 575 mph, faster than many of the first jet aircraft and comparable to jet cruising speeds for most missions. The Bear would serve as their most successful long-range combat and surveillance aircraft and symbol of Soviet power projection throughout the end of the 20th century. The USA would incorporate contra-rotating turboprop engines, such as the ill-fated twin-turbine Allison T40 — essentially a twinned up pair of Allison T38 turboprop engines driving contra-rotating propellers — into a series of experimental aircraft during the 1950s, with aircraft powered with the T40, like the Convair R3Y Tradewind flying boat never entering U.S. Navy service.

The first American turboprop engine was the General Electric XT31, first used in the experimental Consolidated Vultee XP-81. The XP-81 first flew in December 1945, the first aircraft to use a combination of turboprop and turbojet power. The technology of the Allison's earlier T38 design evolved into the Allison T56, with quartets of the T56s being used to power the Lockheed Electra airliner, its military maritime patrol derivative the P-3 Orion, and the widely produced C-130 Hercules military transport aircraft. One of the most produced turboprop engines used in civil aviation is the Pratt & Whitney Canada PT6 engine.

The first turbine-powered, shaft-driven helicopter was the Kaman K-225, a development of Charles Kaman's K-125 synchropter, which used a Boeing T50 turboshaft engine to power it on 11 December 1951.

Current Engines

manufacturer	Country	designation	dry weight (kg)	takeoff rating (kW)	Application
DEMC	People's Republic of China	WJ5E	720	2130	Harbin SH-5, Xi'an Y-7
Europrop International	European Union	TP400-D6	1800	8203	Airbus A400M
General Electric	United States	CT7-5A	365	1294	

manufacturer	Country	designation	dry weight (kg)	takeoff rating (kW)	Application
General Electric	United States	CT7-9	365	1447	CASA/IPTN CN-235, Let L-610, Saab 340, Sukhoi Su-80
General Electric	United States Czech Republic	H80 Series	200	550 - 625	Thrush Model 510, Let 410NG, Let L-410 Turbolet UVP-E, CAIGA Primus 150, Nextant G90XT
General Electric	United States	T64-P4D	538	2535	Aeritalia G.222, de Havilland Canada DHC-5 Buffalo, Kawasaki P-2J
Honeywell	United States	TPE331 Series	150 - 275	478 - 1650	Aero/Rockwell Turbo Commander 680/690/840/960/1000, Antonov An-38, Ayres Thrush, BAe Jetstream 31/32, BAe Jetstream 41, CASA C-212 Aviocar, Cessna 441 Conquest II, Dornier Do 228, Fairchild Swearingen Metroliner, General Atomics MQ-9 Reaper, GrumGeman, Mitsubishi MU-2, North American Rockwell OV-10 Bronco, RUAG Do 228NG, Short SC.7 Skyvan, Short Tucano, Swearingen Merlin, Fairchild Swearingen Metroliner
Honeywell	United States	LTP 101-700	147	522	Air Tractor AT-302, Piaggio P.166
KKBM	Russia	NK-12MV	1900	11033	Antonov An-22, Tupolev Tu-95, Tupolev Tu-114
Progress	Ukraine	TV3-117VMA-SB2	560	1864	Antonov An-140
Klimov	Russia	TV7-117S	530	2100	Ilyushin Il-112, Ilyushin Il-114
Progress	Ukraine	AI20M	1040	2940	Antonov An-12, Antonov An-32, Ilyushin Il-18
Progress	Ukraine	AI24T	600	1880	Antonov An-24, Antonov An-26, Antonov An-30

manufacturer	Country	designation	dry weight (kg)	takeoff rating (kW)	Application
LHTEC	United States	LHTEC T800	517	2013	AgustaWestland Super Lynx 300 (CTS800-4N), AgustaWestland AW159 Lynx Wildcat (CTS800-4N), Ayres LM200 Loadmaster (LHTEC CTP800-4T) (aircraft not built), Sikorsky X2 (T800-LHT-801), TAI/AgustaWestland T-129 (CTS800-4A)
OMKB	Russia	TVD-20	240	1081	Antonov An-3, Antonov An-38
Pratt & Whitney Canada	Canada	PT-6 Series	149 - 260	430 - 1500	Air Tractor AT-502, Air Tractor AT-602, Air Tractor AT-802, Beechcraft Model 99, Beechcraft King Air, Beechcraft Super King Air, Beechcraft 1900, Beechcraft T-6 Texan II, Cessna 208 Caravan, Cessna 425 Corsair/Conquest I, de Havilland Canada DHC-6 Twin Otter, Harbin Y-12, Embraer EMB 110 Bandeirante, Let L-410 Turbolet, Piaggio P.180 Avanti, Pilatus PC-12, Piper PA-42 Cheyenne, Piper PA-46-500TP Meridian, Shorts 360, Daher TBM 700, Daher TBM 850, Daher TBM 900
Pratt & Whitney Canada	Canada	PW120	418	1491	ATR 42-300/320
Pratt & Whitney Canada	Canada	PW121	425	1603	ATR 42-300/320, Bombardier Dash 8 Q100
Pratt & Whitney Canada	Canada	PW123 C/D	450	1603	Bombardier Dash 8 Q300
Pratt & Whitney Canada	Canada	PW127	481	2051	ATR 72
Pratt & Whitney Canada	Canada	PW150A	717	3781	Bombardier Dash 8 Q400
PZL	Poland	TWD-10B	230	754	PZL M28
RKBM	Russia	TVD-1500S	240	1044	Sukhoi Su-80
Rolls-Royce	United Kingdom	Dart Mk 536	569	1700	Avro 748, Fokker F27, Vickers Viscount
Rolls-Royce	United Kingdom	Tyne 21	569	4500	Aeritalia G.222, Breguet Atlantic, Transall C-160

manufacturer	Country	designation	dry weight (kg)	takeoff rating (kW)	Application
Rolls-Royce	🇬🇧 United Kingdom	250-B17	88.4	313	Fuji T-7, Britten-Norman Turbine Islander, O&N Cessna 210, Soloy Cessna 206, Propjet Bonanza
Rolls-Royce	🇬🇧 United Kingdom	T56-14		3433	P-3 Orion
Rolls-Royce	🇬🇧 United Kingdom	T56-15	828	3424	C-130 Hercules
Rolls-Royce	🇬🇧 United Kingdom	T56-27	880	3910	E-2 Hawkeye
Rolls-Royce	🇬🇧 United Kingdom	AE2100A	715.8	3095	Saab 2000
Rolls-Royce	🇬🇧 United Kingdom	AE2100J	710	3424	ShinMaywa US-2
Rolls-Royce	🇬🇧 United Kingdom	AE2100D2, D3	702	3424	Alenia C-27J Spartan, Lockheed Martin C-130J Super Hercules
Rybinsk	🇷🇺 Russia	TVD-1500V	220	1156	
Saturn	🇷🇺 Russia	TAL-34-1	178	809	
Turbomeca	🇫🇷 France	Arrius 1D	111	313	Socata TB 31 Omega
Turbomeca	🇫🇷 France	Arrius 2F	103	376	
Walter	🇨🇿 Czech Republic	M601 Series	200	560	Let L-410 Turbolet, Aerocomp Comp Air 10 XL, Aerocomp Comp Air 7, Ayres Thrush, Dornier Do 28, Lancair Propjet, Let Z-37T, Let L-420, Myasishchev M-101T, PAC FU-24 Fletcher, Progress Rysachok, PZL-106 Kruk, PZL-130 Orlik, SM-92T Turbo Finist
Walter	🇨🇿 Czech Republic	M602A	570	1360	Let L-610
Walter	🇨🇿 Czech Republic	M602B	480	1500	

Auxiliary Power Unit

The APU exhaust at the tail end of an Airbus A380

An auxiliary power unit (APU) is a device on a vehicle that provides energy for functions other than propulsion. They are commonly found on large aircraft and naval ships as well as some large land vehicles. Aircraft APUs generally produce 115 V alternating current (AC) at 400 Hz (rather than 50/60 Hz in mains supply), to run the electrical systems of the aircraft; others can produce 28 V direct current (DC). APUs can provide power through single- or three-phase systems.

Transport Aircraft

Function

The primary purpose of an aircraft APU is to provide power to start the main engines. Turbine engines must be accelerated to a high rotational speed to provide sufficient air compression for self-sustaining operation. Smaller jet engines are usually started by an electric motor, while larger engines are usually started by an air turbine motor. Before the engines are to be turned, the APU is started, generally by a battery or hydraulic accumulator. Once the APU is running, it provides power (electric, pneumatic, or hydraulic, depending on the design) to start the aircraft's main engines.

APIC APS3200 APU for Airbus A320 family.

To start, a jet engine requires pneumatic rotation of the turbine, AC-electrical fuel pumps, and an AC-electrical "flash" that ignites the fuel. As the turbine (behind the combustion chamber) is already rotating, the front inlet fans are also rotating. After the ignition, both fans and turbine speed up their rotation. As combustion stabilizes, the engine thereafter only needs the fuel to run at idle. The started engine can now replace the APU when starting up further engines. During flight the APU and its generator are not needed.

APUs are also used to run accessories while the engines are shut down. This allows the cabin to be comfortable while the passengers are boarding before the aircraft's engines are started. Electrical power is used to run systems for preflight checks. Some APUs are also connected to a hydraulic pump, allowing crews to operate hydraulic equipment (such as flight controls or flaps) prior to engine start. This function can also be used, on some aircraft, as a backup in flight in case of engine or hydraulic failure.

Aircraft with APUs can also accept electrical and pneumatic power from ground equipment when an APU has failed or is not to be used. Some airports reduce the use of APUs due to noise and pollution, and ground power is used when possible.

APUs fitted to extended-range twin-engine operations (ETOPS) aircraft are a critical safety device, as they supply backup electricity and compressed air in place of the dead engine or failed main engine generator. While some APUs may not be startable in flight, ETOPS-compliant APUs must be flight-startable at altitudes up to the aircraft service ceiling. Recent applications have specified starting up to 43,000 ft (13,000 m) from a complete cold-soak condition such as the Hamilton Sundstrand APS5000 for the Boeing 787 Dreamliner. If the APU or its electrical generator is not available, the aircraft cannot be released for ETOPS flight and is forced to take a longer non-ETOPS route.

APUs providing electricity at 400 Hz are smaller and lighter than their 50/60 Hz counterparts, but are costlier; the drawback being that such high frequency systems suffer from voltage drops.

History

The Riedel 2-stroke engine used as the pioneering example of an APU, to turn over the central shaft of both World War II-era German BMW 003 and Junkers Jumo 004 jet engines.

During World War I, the British Coastal class blimps, one of several types of airship operated by the Royal Navy, carried a 1.75 horsepower (1.30 kW) ABC auxiliary engine. These powered a generator for the craft's radio transmitter and, in an emergency, could power an auxiliary air blower. One of the first military fixed-wing aircraft to use an APU was the British, World War 1, Supermarine Nighthawk, an anti-Zeppelin Night fighter.

The Riedel APU unit installed on a preserved BMW 003 jet engine.

During World War 2, a number of large American military aircraft were fitted with APUs. These were typically known as *putt–putts*, even in official training documents. The putt-putt on the B-29 Superfortress bomber was fitted in the unpressurised section at the rear of the aircraft. Various models of four-stroke, Flat-twin or V-twin engines were used. The 7 horsepower (5.2 kW) engine drove a *P2*, DC generator, rated 28.5 Volts and 200 Amps (several of the same *P2* generators, driven by the main engines, were the B-29's DC power source in flight). The putt-putt provided power for starting the main engines and was used after take-off to a height of 10,000 feet (3,000 m). The putt-putt was restarted when the B-29 was descending to land.

Some models of the B-24 Liberator had a putt–putt fitted at the front of the aircraft, inside the nose-wheel compartment. Some models of the Douglas C-47 Skytrain transport aircraft carried a putt-putt under the cockpit floor.

The first German jet engines built during the Second World War used a mechanical APU starting system designed by the German engineer Norbert Riedel. It consisted of a 10 horsepower (7.5 kW) two-stroke flat engine, which for the Junkers Jumo 004 design was hidden in the intake diverter, essentially functioning as a pioneering example of an auxiliary power unit for starting a jet engine. A hole in the extreme nose of the centrebody contained a manual pull-handle which started the piston engine, which in turn rotated the compressor. Two small petrol tanks were fitted in the annular intake. The engine was considered an extreme short stroke (bore / stroke: 70 mm / 35 mm = 2:1) design so it could fit in the hub of the turbine compressor. For reduction it had an integrated planetary gear. It was produced in Victoria in Nuremberg and served as a mechanical APU-style starter for all three German jet engine designs to have made it to at least the prototype stage before May of 1945: the Junkers Jumo 004, the BMW 003, and the prototypes (19 built) of the more advanced Heinkel HeS 011 engine, which mounted it just above the intake passage in the Heinkel-crafted sheetmetal of the engine nacelle nose.

The Boeing 727 in 1963 was the first jetliner to feature a gas turbine APU, allowing it to operate at smaller airports, independent from ground facilities. The APU can be identified on many modern airliners by an exhaust pipe at the aircraft's tail.

Sections

A typical gas turbine APU for commercial transport aircraft comprises three main sections:

Power Section

The power section is the gas generator portion of the engine and produces all the shaft power for the APU.Load Compressor Section

The load compressor is generally a shaft-mounted compressor that provides pneumatic power for the aircraft, though some APUs extract bleed air from the power section compressor. There are two actuated devices: the inlet guide vanes that regulate airflow to the load compressor and the surge control valve that maintains stable or surge-free operation of the turbo machine.

Load Compressor Section

The load compressor is generally a shaft-mounted compressor that provides pneumatic power for the aircraft, though some APUs extract bleed air from the power section compressor. There are two actuated devices: the inlet guide vanes that regulate airflow to the load compressor and the surge control valve that maintains stable or surge-free operation of the turbo machine

Gearbox Section

The gearbox transfers power from the main shaft of the engine to an oil-cooled generator for electrical power. Within the gearbox, power is also transferred to engine accessories such as the fuel control unit, the lubrication module, and cooling fan. In addition, there is also a starter motor connected through the gear train to perform the starting function of the APU. Some APU designs use a combination starter/generator for APU starting and electrical power generation to reduce complexity.

On the Boeing 787 more-electric aircraft, the APU delivers only electricity to the aircraft. The absence of a pneumatic system simplifies the design, but high demand for electricity requires heavier generators.

Onboard solid oxide fuel cell (SOFC) APUs are being researched.

Manufacturers

Two main corporations compete in the aircraft APU market: United Technologies Corporation (through its subsidiaries Pratt & Whitney Canada and Pratt & Whitney AeroPower), and Honeywell International Inc.

Military Aircraft

Smaller military aircraft, such as fighters and attack aircraft, feature auxiliary power systems which are different from those used in transport aircraft. The functions of engine starting and providing electrical and hydraulic power are divided between two units, the *jet fuel starter* and the *emergency power unit*.

Jet Fuel Starter

A jet fuel starter (JFS) is a small turboshaft engine designed to drive a jet engine to its self-accelerating RPM. Rather than supplying bleed air to a starter motor in the manner of an APU, a JFS output shaft is mechanically connected to an engine. As soon as the JFS begins to turn, the engine turns; unlike APUs, these starters are not designed to produce electrical power when engines are not running.

Jet fuel starters use a free power turbine section, but the method of connecting it to the engine depends on the aircraft design. In single-engine aircraft such as the A-7 Corsair II and F-16 Fighting Falcon, the JFS power section is always connected to the main engine through the engine's accessory gearbox. In contrast, the twin-engine F-15 Eagle features a single JFS, and the JFS power section is connected through a central gearbox which can be engaged to one engine at a time. On the F-15, the jet fuel starter (JFS) is mated with a central gearbox (CGB). The CGB has extendable pawl shafts that extend out to reach the aircraft mounted accessory drive (AMAD) mounted in front of each engine. The AMAD is connected to the jet engine by the power takeoff (PTO) shaft. As the engine accelerates to starting speed, the PTO shaft becomes the method to drive the AMAD during flight. Once the aircraft engine has started and begins driving the AMAD, the pawl shaft on the CGB returns to its retracted position and the JFS is shut down.

Emergency Power Unit

Emergency hydraulic and electric power are provided by a different type of gas turbine engine. Unlike most gas turbines, an emergency power unit has no gas compressor or ignitors, and uses a combination of hydrazine and water, rather than jet fuel. When the hydrazine and water mixture is released and passes across a catalyst of iridium, it spontaneously ignites, creating hot expanding gases which drive the turbine. The power created is transmitted through a gearbox to drive an electrical generator and hydraulic pump.

The hydrazine is contained in a sealed, nitrogen charged accumulator. When the system is armed, the hydrazine is released whenever the engine-driven generators go off-line, or if all engine-driven hydraulic pumps fail.

Airport Equipment

Many airports have adopted APUs as a solution to high fuel consumption. Much of the equipment used to clean and clear runways will use an average of two or more gallons per hour of diesel while idling. Adding an APU will provide power, heating and cooling as well as hydraulic warming if necessary and can result in significant fuel and maintenance savings.

This HP2000 brand APU is mounted on a runway sweeper at O'Hare airport in Chicago IL.

Spacecraft

The Space Shuttle APUs provided hydraulic pressure. The Space Shuttle had three redundant APUs, powered by hydrazine fuel. They were only powered up for ascent, re-entry, and landing. During ascent, the APUs provided hydraulic power for gimballing of the Shuttle's three engines and control of their large valves, and for movement of the control surfaces. During landing, they moved the control surfaces, lowered the wheels, and powered the brakes and nose-wheel steering. Landing could be accomplished with only one APU working. In the early years of the Shuttle there were problems with APU reliability, with malfunctions on three of the first nine Shuttle missions.

Armor

APUs are fitted to some tanks to provide electrical power without the high fuel consumption and large infrared signature of the main engine. Both the M1 Abrams and variants of the Leopard 2 such as the Spanish and Danish variants carry the APU in the rear right hull section. The British Centurion tank used an Austin A-Series inline-4 as its auxiliary power unit. The Turkish self-propelled howitzer T-155 Fırtına uses a 2-stroke diesel engine located at the rear right hull to supply power to fire control computers and turret hydraulics.

Towed Artillery

Many modern towed artillery pieces are fitted with internal combustion engines, primarily to provide hydraulic power to aid in gun laying and to power flick rammers or other aids to loading. These engines can also be used to provide limited battlefield mobility when no artillery tractors are available.

Commercial Vehicles

A refrigerated or frozen food semi trailer or train car may be equipped with an independent APU and fuel tank to maintain low temperatures while in transit, without the need for an external transport-supplied power source.

In the United States, federal Department of Transportation regulations require 10 hours of rest for every 11 hours of driving. When stopped, drivers often idle their engines to provide heat, light, and power. Idling inefficiently burns fuel and puts wear on engines. Some trucks carry an APU designed to eliminate these long idles. An APU can save up to 20 US gallons (76 L) (Cat 600 – 10 hours downtime @ 2 gallons per hour idling) of fuel a day, and can extend the useful life of the main engine by around 100,000 miles (160,000 km), by reducing non-productive run time.

On some older diesel engines, an APU was used instead of an electric motor to start the main engine. These were primarily used on large pieces of construction equipment.

Diesel

The most common APU for a commercial truck is a small diesel engine with its own cooling system, heating system, generator or alternator system with or without inverter, and air conditioning compressor, housed in an enclosure and mounted to one of the frame rails of a semi-truck. Other designs fully integrate the auxiliary cooling, heating, and electrical components throughout the chassis of the truck. The APU generator engine is a fraction of the main engine's size and uses a fraction of the fuel; some models can run for eight hours on 1 US gallon (3.8 L) of diesel. The generator also powers the main engine's block and fuel system heaters, so the main engine can be started easily right before departure if the APU is allowed to run for a period beforehand. These units are used to provide climate control and electrical power for the truck's sleeper cab and engine block heater during downtime on the road as mandated by statewide laws for idle reduction.

This is the inside of an HP2000 Auxiliary Power unit (APU) surrounded by heat pump components.

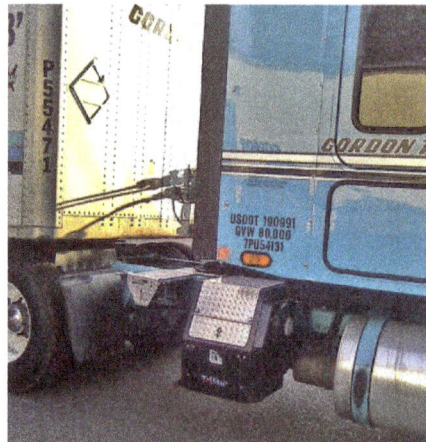

Diesel-powered APU on truck

Propane

A less common APU available for commercial diesel truck cabs and sleepers includes a heating & cooling system, duplex 110 volt electrical power outlets inside and outside the cab. These are all powered by a single propane fueled generator. The Tri Pac III APU was introduced to the market in 2012 by American Truck Group, LLC. With this system in place, it is no longer necessary to idle the truck engine during rest and sleep periods in order to provide the air condition/heating and appliance power for the driver. This will reduce non travel operating hours on the engine and extend its useful mileage by up to an estimated 100,000 miles.

Electric

An electric APU installed on a truck

Electric APUs have started gaining acceptance. These electric APUs use battery packs instead of the diesel engine on traditional APUs as a source of power. The APU's battery pack is charged when the truck is in motion. When the truck is idle, the stored energy in the battery pack is then used to power an air conditioner, heater, and other devices (television, microwave oven, etc.) in the bunk.

Fuel Cells

In recent years, truck and fuel cell manufacturers have teamed up to create, test and demonstrate a fuel cell APU that eliminates nearly all emissions and uses diesel fuel more efficiently. In 2008, a DOE sponsored partnership between Delphi Electronics and Peterbilt demonstrated that a fuel cell could provide power to the electronics and air conditioning of a Peterbilt Model 386 under simulated "idling" conditions for 10 hours. Delphi has said the 5 kW system for Class 8 trucks will be released in 2012, at an $8000–9000 price tag that would be competitive with other "midrange" two-cylinder diesel APUs, should they be able to meet those deadlines and cost estimates.

Other Forms of Transport

Where the elimination of exhaust emissions or noise is particularly important (such as yachts, camper vans), fuel cells and photovoltaic modules are used as APUs for electricity generation.

APUs are also installed on some diesel locomotives, allowing the prime mover to be shut down during extended idle periods, while providing power and heat to maintain air pressure and keep the batteries charged and the engine coolant water from freezing.

Turboshaft

A turboshaft engine is a form of gas turbine which is optimized to produce shaft power rather than jet thrust.

In concept, turboshaft engines are very similar to turbojets, with additional turbine expansion to extract heat energy from the exhaust and convert it into output shaft power. They are even more similar to turboprops, with only minor differences, and a single engine is often sold in both forms.

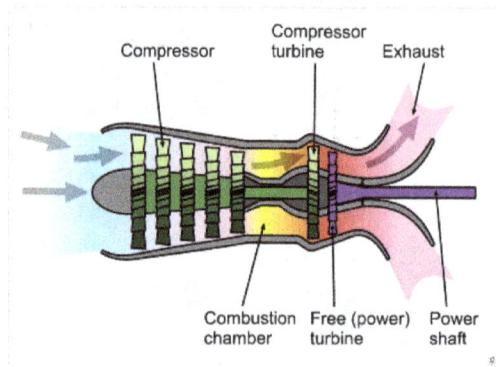

Schematic diagram showing the operation of a simplified turboshaft engine. The compressor spool is shown in green and the free / power spool is in purple.

Turboshaft engines are commonly used in applications that require a sustained high power output, high reliability, small size, and light weight. These include helicopters, auxiliary power units, boats and ships, tanks, hovercraft, and stationary equipment.

Overview

A turboshaft engine may be made up of two major parts assemblies: the 'gas generator' and the 'power section'. The gas generator consists of the compressor, combustion chambers with ignitors and fuel nozzles, and one or more stages of turbine. The power section consists of additional stages of turbines, a gear reduction system, and the shaft output. The gas generator creates the hot expanding gases to drive the power section. Depending on the design, the engine accessories may be driven either by the gas generator or by the power section.

In most designs, the gas generator and power section are mechanically separate so they can each rotate at different speeds appropriate for the conditions, referred to as a 'free power turbine'. A free power turbine can be an extremely useful design feature for vehicles, as it allows the design to forgo the weight and cost of complex multiple-ratio transmissions and clutches.

The general layout of a turboshaft is similar to that of a turboprop. The main difference is a turboprop is structurally designed to support the loads created by a rotating propeller, as the propeller is not attached to anything but the engine itself. In contrast, turboshaft engines usually drive a transmission which is not structurally attached to the engine. The transmission is attached to the vehicle structure and supports the loads created instead of the engine. In practice, though, many of the same engines are built in both turboprop and turboshaft versions, with only minor differences.

An unusual example of the turboshaft principle is the Pratt & Whitney F135-PW-600 turbofan engine for the STOVL F-35B – in conventional mode it operates as a turbofan, but when powering the LiftFan, it switches partially to turboshaft mode to send 29,000 horsepower forward through a shaft (like a turboprop) and partially to turbofan mode to continue to send thrust to the main engine's fan and rear nozzle.

Large helicopters use two or three turboshaft engines for redundancy. The Mil Mi-26 uses two

Lotarev D-136 at 11,400 hp each, while the Sikorsky CH-53E Super Stallion uses three General Electric T64 at 4,380 hp each.

Early turboshaft engines were adaptations of turboprop engines, delivering power through a shaft driven directly from the gas generator shafts, via a reduction gearbox. Examples of direct-drive turboshafts include marinised or industrial Rolls-Royce Dart engines.

History

The first examples of a gas turbine engine design ever considered for armoured fighting vehicles, the BMW 003-based GT 101, were tested in Nazi Germany's Panther tanks in mid-1944. The first true turboshaft engine for helicopters was built by the French engine firm Turbomeca, led by the founder, Joseph Szydlowski. In 1948, they built the first French-designed turbine engine, the 100-shp 782. Originally conceived as an auxiliary power unit, it was soon adapted to aircraft propulsion, and found a niche as a powerplant for turboshaft-driven helicopters in the 1950s. In 1950, this work was used to develop the larger 280-shp Artouste, which was widely used on the Aérospatiale Alouette II and other helicopters. This was following the experimental installation of a Boeing T50 turboshaft in an example of the Kaman K-225 synchropter on December 11, 1951, as the world's first-ever turboshaft-powered helicopter of any type to fly. The T-80 tank entered service with the Soviet Army in 1976, that was the first tank using gas turbine for the main engine. Since 1980 the US Army has operated the M1 Abrams tank. Designed and built in the late 1970s, this tank makes use of a gas turbine engine compared to most tanks using reciprocating diesels. The engine has considerably fewer parts, mechanically is very reliable, produces reduced exterior noise, and runs on virtually any fuel: petrol (gasoline), diesel fuel, aviation fuels. Similarly, the Swedish Stridsvagn 103 was the first tank to utilize a gas turbine, in this case as a secondary, high-horsepower "sprint" engine to augment the primary piston engine's performance.

Austin 250hp gas turbine, sectioned

Combined Cycle

In electric power generation a combined cycle is an assembly of heat engines that work in tandem from the same source of heat, converting it into mechanical energy, which in turn usually drives electrical generators. The principle is that after completing its cycle (in the first engine), the working fluid of the first heat engine is still low enough in its entropy that a second subsequent heat engine may extract energy from the waste heat (energy) of the working fluid of the first engine. By combining these multiple streams of work upon a single mechanical shaft turning an electric generator, the overall net efficiency of the system may be increased by 50 – 60 percent. That is, from an overall efficiency of say 34% (in a single cycle) to possibly an overall efficiency of 51% (in a mechanical combination of two cycles) in net Carnot thermodynamic efficiency. This can be done because heat engines are only able to use a portion of the energy their fuel generates (usually less than 50%). In an ordinary (non combined cycle) heat engine the remaining heat (e.g., hot exhaust fumes) from combustion is generally wasted.

Combining two or more thermodynamic cycles results in improved overall efficiency, reducing fuel costs. In stationary power plants, a widely used combination is a gas turbine (operating by the Brayton cycle) burning natural gas or synthesis gas from coal, whose hot exhaust powers a steam power plant (operating by the Rankine cycle). This is called a Combined Cycle Gas Turbine (CCGT) plant, and can achieve a best-of-class real thermal efficiency of around 54% in base-load operation, in contrast to a single cycle steam power plant which is limited to efficiencies of around 35-42%. Many new gas power plants in North America and Europe are of the Combined Cycle Gas Turbine type. Such an arrangement is also used for marine propulsion, and is called a *combined gas and steam (COGAS)* plant. Multiple stage turbine or steam cycles are also common.

Other historically successful combined cycles have used hot cycles with mercury vapor turbines, magnetohydrodynamic generators or molten carbonate fuel cells, with steam plants for the low temperature "bottoming" cycle. Bottoming cycles operating from a steam condenser's heat exhaust are theoretically possible, but uneconomical because of the very large, expensive equipment needed to extract energy from the small temperature differences between condensing steam and outside air or water. However, it is common in cold climates (such as Finland) to drive community heating systems from a power plant's condenser heat. Such cogeneration systems can yield theoretical efficiencies above 95%.

In automotive and aeronautical engines, turbines have been driven from the exhausts of Otto and Diesel cycles. These are called turbo-compound engines.

Basic Combined Cycle

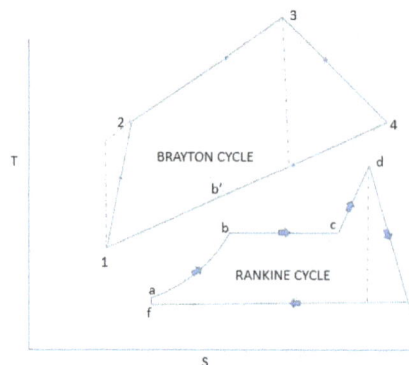

Topping and bottoming cycles

The thermodynamic cycle of the basic combined cycle consists of two power plant cycles. One is the Joule or Brayton cycle which is a gas turbine cycle and the other is Rankine cycle which is a steam turbine cycle. The cycle 1-2-3-4-1 which is the gas turbine power plant cycle is the topping cycle. It depicts the heat and work transfer process taking place in high temperature region.

The cycle a-b-c-d-e-f-a which is the Rankine steam cycle takes place at a low temperature and is known as the bottoming cycle. Transfer of heat energy from high temperature exhaust gas to water and steam takes place by a waste heat recovery boiler in the bottoming cycle. During the constant pressure process 4-1 the exhaust gases in the gas turbine reject heat. The feed water, wet and super heated steam absorb some of this heat in the process a-b, b-c and c-d.

Steam Generators

The steam power plant gets its input heat from the high temperature exhaust gases from gas turbine power plant. The steam generated thus can be used to drive steam turbine. The Waste Heat Recovery Boiler (WHRB) has 3 sections: Economiser, evaporator and superheater.

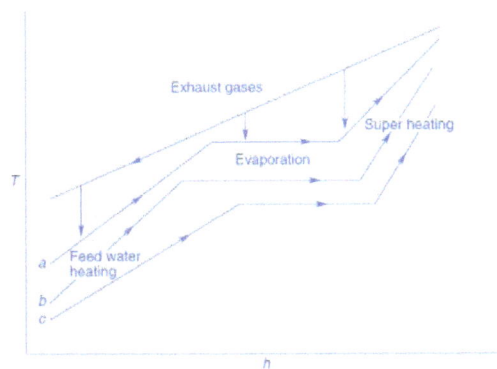

Heat transfer from hot gases to water and steam

Design Principle

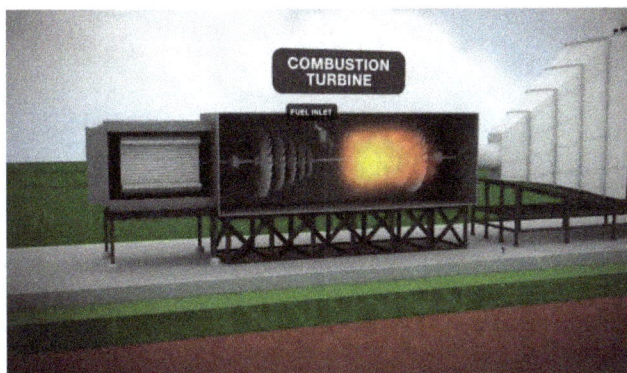

Explanation of the layout and principle of a combined cycle power generator.

The efficiency of a heat engine, the fraction of input heat energy that can be converted to useful work, is limited by the temperature difference between the heat entering the engine and the exhaust heat leaving the engine.

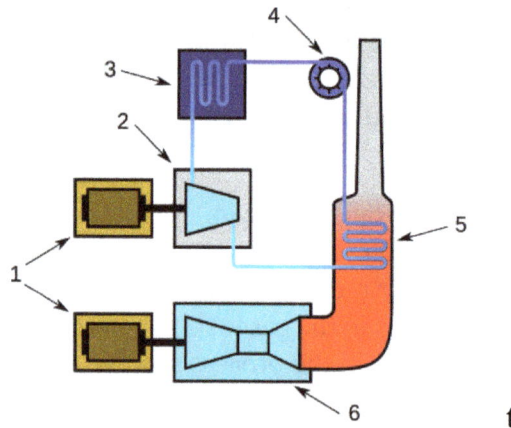

Working principle of a combined cycle power plant (Legend: 1-Electric generators, 2-Steam turbine, 3-Condenser, 4-Pump, 5-Boiler/heat exchanger, 6-Gas turbine)

In a thermal power station, water is the working medium. High pressure steam requires strong, bulky components. High temperatures require expensive alloys made from nickel or cobalt, rather than inexpensive steel. These alloys limit practical steam temperatures to 655 °C while the lower temperature of a steam plant is fixed by the temperature of the cooling water. With these limits, a steam plant has a fixed upper efficiency of 35% to 42%.

An open circuit gas turbine cycle has a compressor, a combustor and a turbine. For gas turbines the amount of metal that must withstand the high temperatures and pressures is small, and lower quantities of expensive materials can be used. In this type of cycle, the input temperature to the turbine (the firing temperature), is relatively high (900 to 1,400 °C). The output temperature of the flue gas is also high (450 to 650 °C). This is therefore high enough to provide heat for a second cycle which uses steam as the working fluid (a Rankine cycle).

In a combined cycle power plant, the heat of the gas turbine's exhaust is used to generate steam by passing it through a heat recovery steam generator (HRSG) with a live steam temperature between 420 and 580 °C. The condenser of the Rankine cycle is usually cooled by water from a lake, river, sea or cooling towers. This temperature can be as low as 15 °C.

Typical Size of CCGT Plants

For large-scale power generation, a typical set would be a 270 MW primary gas turbine coupled to a 130 MW secondary steam turbine, giving a total output of 400 MW. A typical power station might consist of between 1 and 6 such sets.

Plant size is important in the cost of the plant. The larger plant sizes benefit from economies of scale (lower initial cost per kilowatt) and improved efficiency.

Gas turbines of about 150 MW size are already in operation manufactured by at least four separate groups – General Electric and its licensees, Alstom, Siemens, and Westinghouse/Mitsubishi. These groups are also developing, testing and/or marketing gas turbine sizes of about 200 MW. Combined cycle units are made up of one or more such gas turbines, each with a waste heat steam generator arranged to supply steam to a single steam turbine, thus forming a combined cycle block or unit.

Typical Combined cycle block sizes offered by three major manufacturers (Alstom, General Electric and Siemens) are roughly in the range of 50 MW to 500 MW and costs are about $600/kW.

Unfired Boiler

The heat recovery boiler is item 5 in the COGAS figure shown above. No combustion of fuel means that there is no need of a fuel handling plant, and it is simply a heat exchanger. Exhaust enters in super heater and the evaporator and then in to economiser section as it flows out from the boiler. Feed water comes in through the economizer and then exits after having attained saturation temp in the water or steam circuit. Finally it then flows through evaporator and super heater. If the temperature of the gases entering the heat recovery boiler is higher, then the temperature of the exiting gases is also high.

Dual Pressure boiler

It is often desirable if high heat is recovered from the exiting gases. Hence dual pressure boiler is employed for this purpose. It has two water/steam drums. Low pressure drum is connected to low pressure economizer or evaporator. The low pressure steam is generated in low temperature zone. The low pressure steam is supplied to the low temperature turbine. Super heater can be provided in the low pressure circuit.

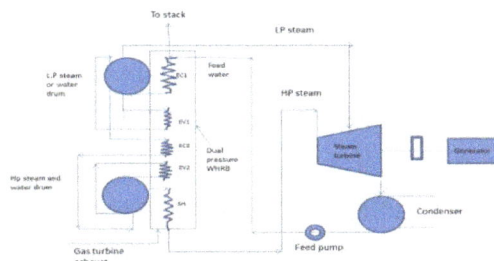

Steam turbine plant lay out with dual pressure heat recovery boiler

Some part of the feed water from the low-pressure zone is transferred to the high-pressure economizer by a booster pump. This economizer heats up the water to its saturation temperature. This saturated water goes through the high-temperature zone of the boiler and is supplied to the high-pressure turbine.

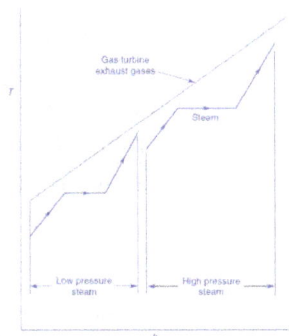

Heat exchange in dual pressure heat recovery boiler

Supplementary Firing

Supplementary firing may be used in combined cycles (in the HRSG) raising exhaust temperatures from 600 °C (GT exhaust) to 800 or even 1000 °C. Using supplemental firing will however not raise the combined cycle efficiency for most combined cycles. For single boilers it may raise the efficiency if fired to 700–750 °C; for multiple boilers however, supplemental firing is often used to improve peak power production of the unit, or to enable higher steam production to compensate for failure of a second unit.

Maximum supplementary firing refers to the maximum fuel that can be fired with the oxygen available in the gas turbine exhaust. The steam cycle is conventional with reheat and regeneration. Hot gas turbine exhaust is used as the combustion air. Regenerative air preheater is not required. A fresh air fan which makes it possible to operate the steam plant even when the gas turbine is not in operation, increases the availability of the unit.

The use of large supplementary firing in Combined Cycle Systems with high gas turbine inlet temperatures causes the efficiency to drop. For this reason the Combined Cycle Plants with maximum supplementary firing are only of minimal importance today, in comparison to simple Combined Cycle installations. However, they have two advantages that is a) coal can be burned in the steam generator as the supplementary fuel, b) has very good part load efficiency.

The HRSG can be designed with supplementary firing of fuel after the gas turbine in order to increase the quantity or temperature of the steam generated. Without supplementary firing, the efficiency of the combined cycle power plant is higher, but supplementary firing lets the plant respond to fluctuations of electrical load. Supplementary burners are also called *duct burners*.

More fuel is sometimes added to the turbine's exhaust. This is possible because the turbine exhaust gas (flue gas) still contains some oxygen. Temperature limits at the gas turbine inlet force the turbine to use excess air, above the optimal stoichiometric ratio to burn the fuel. Often in gas turbine designs part of the compressed air flow bypasses the burner and is used to cool the turbine blades.

Supplementary firing raises the temperature of the exhaust gas from 800 to 900 degree Celsius. Relatively high flue gas temperature raises the condition of steam (84 bar, 525 degree Celsius) thereby improving the efficiency of steam cycle.

Fuel for Combined Cycle Power Plants

Combined cycle plants are usually powered by natural gas, although fuel oil, synthesis gas or other fuels can be used. The supplementary fuel may be natural gas, fuel oil, or coal. Biofuels can also be used. Integrated solar combined cycle power stations combine the energy harvested from solar radiation with another fuel to cut fuel costs and environmental impact (look ISCC section). Next generation nuclear power plants are also on the drawing board which will take advantage of the higher temperature range made available by the Brayton top cycle, as well as the increase in thermal efficiency offered by a Rankine bottoming cycle.

The improvement in shale gas extraction has increased natural gas supplies and reserves dramatically. Because of this fact, it is becoming the fuel of choice for an increasing amount of private investors and consumers because it is more versatile than coal or oil and can be used in 90% of energy applications. Chile which once depended on hydro-power for 70% of its electricity supply, is now boosting its gas supplies to reduce reliance on its drought afflicted hydro dams. Similarly China is tapping its gas reserves to reduce reliance on coal, which is currently burned to generate 80% of the country's electricity supply.

Where the extension of a gas pipeline is impractical or cannot be economically justified, electricity needs in remote areas can be met with small-scale Combined Cycle Plants, using renewable fuels. Instead of natural gas, Combined Cycle Plants can be filled with biogas derived from agricultural and forestry waste, which is often readily available in rural areas.

Low-Grade Fuel for Turbines

Gas turbines burn mainly natural gas and light oil. Crude oil, residual, and some distillates contain corrosive components and as such require fuel treatment equipment. In addition, ash deposits from these fuels result in gas turbine deratings of up to 15 percent. They may still be economically attractive fuels however, particularly in combined-cycle plants.

Sodium and potassium are removed from residual, crude and heavy distillates by a water washing procedure. A simpler and less expensive purification system will do the same job for light crude and light distillates. A magnesium additive system may also be needed to reduce the corrosive effects if vanadium is present. Fuels requiring such treatment must have a separate fuel-treatment plant and a system of accurate fuel monitoring to assure reliable, low-maintenance operation of gas turbines.

Configuration

A single shaft combined cycle plant comprises a gas turbine and a steam turbine driving a common generator. In a multi-shaft combined cycle plant, each gas turbine and each steam turbine has its own generator. The single shaft design provides slightly less initial cost and slightly better efficiency than if the gas and steam turbines had their own generators. The multi-shaft design enables two or more gas turbines to operate in conjunction with a single steam turbine, which can be more economical than a number of single shaft units.

The primary disadvantage of single shaft combined cycle power plants is that the number of steam turbines, condensers and condensate systems – and perhaps the number of cooling towers and circulating water systems – increases to match the number of gas turbines. For a multi-shaft combined cycle power plant there is only one steam turbine, condenser and the rest of the heat sink for up to three gas turbines; only their size increases. Having only one large steam turbine and heat sink results in low cost because of economies of scale. A larger steam turbine also allows the use of higher pressures and results in a more efficient steam cycle. Thus the overall plant size and the associated number of gas turbines required have a major impact on whether a single shaft combined cycle power plant or a multiple shaft combined cycle power plant is more economical.

The combined-cycle system includes single-shaft and multi-shaft configurations. The single-shaft system consists of one gas turbine, one steam turbine, one generator and one Heat Recovery Steam Generator (HRSG), with the gas turbine and steam turbine coupled to the single generator in a tandem arrangement on a single shaft. Key advantages of the single-shaft arrangement are operating simplicity, smaller footprint, and lower startup cost. Single-shaft arrangements, however, will tend to have less flexibility and equivalent reliability than multi-shaft blocks. Additional operational flexibility is provided with a steam turbine which can be disconnected, using a synchro-self-shifting (SSS) Clutch, for start up or for simple cycle operation of the gas turbine.

Multi-shaft systems have one or more gas turbine-generators and HRSGs that supply steam through a common header to a separate single steam turbine-generator. In terms of overall investment a multi-shaft system is about 5% higher in costs.

Single- and multiple-pressure non-reheat steam cycles are applied to combined-cycle systems equipped with gas turbines having rating point exhaust gas temperatures of approximately 540 °C or less. Selection of a single- or multiple-pressure steam cycle for a specific application is determined by economic evaluation which considers plant installed cost, fuel cost and quality, plant duty cycle, and operating and maintenance cost.

Multiple-pressure reheat steam cycles are applied to combined-cycle systems with gas turbines having rating point exhaust gas temperatures of approximately 600 °C.

The most efficient power generation cycles are those with unfired HRSGs with modular pre-engineered components. These unfired steam cycles are also the lowest in cost. Supplementary-fired combined-cycle systems are provided for specific application.

The primary regions of interest for cogeneration combined-cycle systems are those with unfired and supplementary fired steam cycles. These systems provide a wide range of thermal energy to electric power ratio and represent the range of thermal energy capability and power generation covered by the product line for thermal energy and power systems.

Efficiency of CCGT Plants

To avoid confusion, the efficiency of heat engines and power stations should be stated relative to the Higher Heating Value (HHV) or Lower Heating Value (LHV) of the fuel, to include or exclude the heat that can be obtained from condensing the flue gas. It should also be specified whether Gross output at the generator terminals or Net Output at the power station fence is being considered.

The LHV figure is NOT a computation of electricity net energy compared to energy content of fuel input; it is 11% higher than that. The HHV figure is a computation of electricity net energy compared to energy content of fuel input. If the LHV approach were used for some new condensing boilers, the efficiency would calculate to be over 100%. Manufacturers prefer to cite the higher LHV efficiency, e.g. 60%, for a new CCGT, but utilities, when calculated how much electricity the plant will generate, divide this by 1.11 to get the real, e.g. 54%, HHV efficiency of that CCGT. Coal plant efficiencies are computed on a HHV basis (it doesn't make nearly as much difference for coal burn, as for gas).

The difference between HHV and LHV for gas, can be estimated (using USA units) by 1055Btu/Lb * w, where w is the lbs of water after combustion per lb of fuel. To convert the HHV of natural gas, which is 23875 Btu/lb, to an LHV (methane is 25% hydrogen) would be: 23875 − (1055*0.25*18/2) = 21500. Because the efficiency is determined by dividing the energy output by the input, and the input on an LHV basis is smaller than the HHV basis, the overall efficiency on an LHV basis is higher. So, using the ratio: 23875/21500 = 1.11 you can convert the HHV to an LHV.

So a real best-of-class baseload CCGT efficiency of 54%, as experienced by the utility operating the plant, translates to 60% LHV as the manufacturer's published headline CCGT efficiency.

In general in service Combined Cycle efficiencies are over 50 percent on a lower heating value and Gross Output basis. Most combined cycle units, especially the larger units, have peak, steady state efficiencies on the LHV (marketing figure) basis of 55 to 59%. Research aimed at 1370 °C (2500 °F) turbine inlet temperature has led to even more efficient combined cycles and nearly 60 percent LHV efficiency (54% HHV efficiency) has been reached in the combined cycle unit of Baglan Bay, a GE H-technology gas turbine with a NEM 3 pressure reheat boiler, utilising steam from the HRSG to cool the turbine blades.

Siemens AG announced in May 2011 to have achieved a 60.75% net efficiency with a 578 mega-watts SGT5-8000H gas turbine at the Irsching Power Station.

The most recent General Electric 9HA can attain 41.5% simple cycle efficiency and 61.4% in combined cycle mode, with a gas turbine output of 397 to 470MW and a combined output of 592MW to 701MW. Its firing temperature is between 2,600 and 2,900 °F (1,430 and 1,590 °C), its overall pressure ratio is 21.8 to 1 and is scheduled to be used by Électricité de France in Bouchain. On April 28, 2016 this plant was certified by Guinness World Records as the worlds most efficient combined cycle power plant at 62.22%. The Chubu Electric's Nishi-ku, Nagoya power plant 405MW 7HA is expected to have 62% gross combined cycle efficiency.

By combining both gas and steam cycles, high input temperatures and low output temperatures can be achieved. The efficiency of the cycles add, because they are powered by the same fuel source. So, a combined cycle plant has a thermodynamic cycle that operates between the gas-turbine's high firing temperature and the waste heat temperature from the condensers of the steam cycle. This large range means that the Carnot efficiency of the cycle is high. The actual efficiency, while lower than this, is still higher than that of either plant on its own. The actual efficiency achievable is a complex area.

The electric efficiency of a combined cycle power station, if calculated as electric energy produced as a percent of the lower heating value of the fuel consumed, may be as high as 58 percent when operating new, i.e. unaged, and at continuous output which are ideal conditions. As with single cycle thermal units, combined cycle units may also deliver low temperature heat energy for industrial processes, district heating and other uses. This is called cogeneration and such power plants are often referred to as a Combined Heat and Power (CHP) plant.

Boosting Efficiency

The efficiency of CCGT and GT can be boosted by pre-cooling combustion air. This is practised in hot climates and also has the effect of increasing power output. This is achieved by evaporative cooling of water using a moist matrix placed in front of the turbine, or by using Ice storage air conditioning. The latter has the advantage of greater improvements due to the lower temperatures available. Furthermore, ice storage can be used as a means of load control or load shifting since ice can be made during periods of low power demand and, potentially in the future the anticipated high availability of other resources such as renewables during certain periods.

Integrated Gasification Combined Cycle (IGCC)

An integrated gasification combined cycle, or IGCC, is a power plant using synthesis gas (syngas). Syngas can be produced from a number of sources, including coal and biomass. The system utilizes gas and steam turbines, the steam turbine operating off of the heat leftover from the gas turbine. This process can raise electricity generation efficiency to around 50%.

Integrated Solar Combined Cycle (ISCC)

An Integrated Solar Combined Cycle (ISCC) is a hybrid technology in which a solar thermal field is integrated within a combined cycle plant. In ISCC plants, solar energy is used as an auxiliary heat supply, supporting the steam cycle, which results in increased generation capacity or a reduction of fossil fuel use.

Thermodynamic benefits are that daily steam turbine startup losses are eliminated. Major factors limiting the load output of a combined cycle power plant are the allowed pressure and temperature transients of the steam turbine and the heat recovery steam generator waiting times to establish required steam chemistry conditions and warm-up times for the balance of plant and the main piping system. Those limitations also influence the fast start-up capability of the gas turbine by requiring waiting times. And waiting gas turbines consume gas. The solar component, if the plant is started after sunshine, or before, if we have heat storage, allows us to preheat the steam to the required conditions. That is, the plant is started faster and we consume less gas before achieving operating conditions. Economic benefits are that the solar components costs are 25% to 75% those of a Solar Energy Generating Systems plant of the same collector surface.

The first such system to come online was the Archimede solar power plant, Italy in 2010, followed by Martin Next Generation Solar Energy Center in Florida, and in 2011 by the Kuraymat ISCC Power Plant in Egypt, Yazd power plant in Iran, Hassi R'mel in Algeria, Ain Beni Mathar in Morocco.

Automotive Use

Combined cycles have traditionally only been used in large power plants. BMW, however, has proposed that automobiles use exhaust heat to drive steam turbines. This can even be connected to the car or truck's cooling system to save space and weight, but also to provide a condenser in the same location as the radiator and preheating of the water using heat from the engine block.

It may be possible to use the pistons in a reciprocating engine for both combustion and steam expansion as in the Crower engine.

A turbocharged car is also a combined cycle. Bowman of Southampton offer a commercially proven add-on turbocharger which additionally can generate electric power lowering overall fuel consumption by about 8%.

Aeromotive Use

Some versions of the Wright R-3350 were produced as turbo-compound engines. Three turbines driven by exhaust gases, known as *power recovery turbines*, provided nearly 600 hp at takeoff. These turbines added power to the engine crankshaft through bevel gears and fluid couplings.

There have been many successful turbo-compound engine designs particularly for aircraft but their mechanical complexity and weight are less economical than multistage turbine engines. Stirling engines are also a good theoretical fit for this application.

Closed-Cycle Gas Turbine

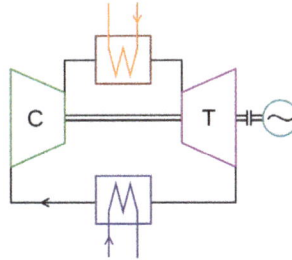

Closed-Cycle Gas Turbine Schematic

C compressor and T turbine assembly

w high-temperature heat exchanger

м low-temperature heat exchanger

~ mechanical load, e.g. electric generator

A closed-cycle gas turbine is a turbine that uses a gas (e.g. air, nitrogen, helium, argon, etc.) for the working fluid as part of a closed thermodynamic system. Heat is supplied from an external source. Such recirculating turbines follow the Brayton cycle.

Background

The initial patent for a closed-cycle gas turbine (CCGT) was issued in 1935 and they were first used commercially in 1939. Seven CCGT units were built in Switzerland and Germany by 1978. Historically, CCGTs found most use as external combustion engines "with fuels such as bituminous coal, brown coal and blast furnace gas" but were superseded by open cycle gas turbines using clean-burning fuels (e.g. "gas or light oil"), especially in highly efficient combined cycle systems. Air-based CCGT systems have demonstrated very high availability and reliability. The most notable helium-based system thus far was *Oberhausen 2*, a 50 megawatt cogeneration plant that operated from 1975 to 1987 in Germany. Compared to Europe where the technology was originally developed, CCGT is not well known in the US.

Nuclear Power

Gas-cooled reactors powering helium-based closed-cycle gas turbines were suggested in 1945. The experimental ML-1 nuclear reactor in the early-1960s used a nitrogen-based CCGT operating at 0.9 MPa. The cancelled pebble bed modular reactor was intended to be coupled with a helium CCGT. Future nuclear (Generation IV reactors) may employ CCGT for power generation, e.g. Flibe Energy intends to produce a liquid fluoride thorium reactor coupled with a CCGT.

Development

Closed-cycle gas turbines hold promise for use with future high temperature solar power and fusion power generation.

They have also been proposed as a technology for use in long-term space exploration.

Supercritical carbon dioxide closed-cycle gas turbines are under development; "The main advantage of the supercritical CO_2 cycle is comparable efficiency with the helium Brayton cycle at significantly lower temperature" (550°C vs. 850°C), but with the disadvantage of higher pressure (20 MPa vs. 8 MPa).

Simple Cycle Combustion Turbine

A Simple-Cycle Combustion Turbine (SCCT) is a type of gas turbine most frequently used in the power generation, aviation (jet engine), and oil and gas industry (electricity generation and mechanical drives). The simple-cycle combustion turbine follows the Brayton Cycle and differs from a combined cycle operation in that it has only one power cycle (ie. no provision for waste heat recovery).

Advantages

There are several advantages of an SCCT. The primary advantage of a SCCT is the high power generated to weight (or size) ratio, when compared to alternatives.

Another advantage is the ability for it to quickly reach full power. Unlike other baseload power plants that may have a minimum time of being online once started. This "minimum up" is a common term in the power industry when referring to this requirement. Therefore SCCTs are usually used as peaking power plants, which can operate from several hours per day to a couple of dozen hours per year, depending on the electricity demand and the generating capacity of the region. In areas with a shortage of base load and load following power plant capacity, a gas turbine power plant may regularly operate during most hours of the day and even into the evening. A typical large simple-cycle gas turbine may produce 100 to 300 megawatts of power and have 35–40% thermal efficiency. The most efficient turbines have reached 46% efficiency.

For power generation applications, the investment costs are cheaper than combined cycle combustion turbine plants (in 2003, the Energy Information Administration estimated that the cost of a combined cycle plant was US$500-550/kW, as opposed to the SCCT cost of US$389/kW), but at reduced efficiency. SCCTs require smaller capital investment than either coal or nuclear power plants and can be scaled to generate small or large amounts of power. Also, the actual construction process can take as little as several weeks to a few months, compared to years for base load power plants.

Disadvantages

A simple cycle combustion turbine has a lower thermal efficiency than a combined cycle machine. Although, they may be less expensive to build simple cycle combustion turbines, due to their low efficiency, cost more to run than most other plants. This results in increased cost per kWh during peak electricity times. Since compared to other NG fuel using turbines, such Combined Cycle Turbines, the lack of efficiency causes the need to increase the amount of NG fuel consumed to produce the same amount of electricity.

Free-Turbine Turboshaft

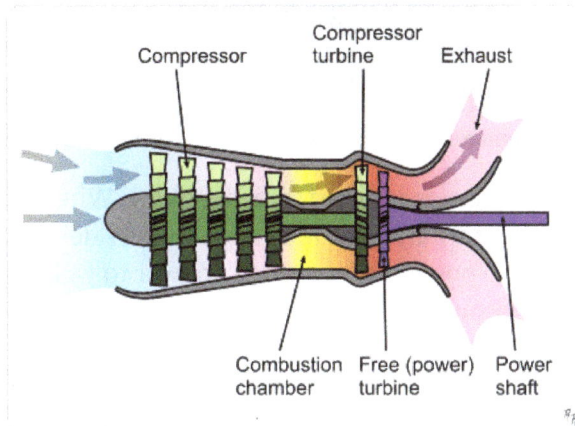

A free-turbine turboshaft is a form of turboshaft or turboprop gas turbine engine where the power is extracted from the exhaust stream of a gas turbine by a separate turbine, downstream of the gas turbine and is not connected to the gas turbine. This is opposed to the power being extracted from the power spool via a gear box.

The advantage of the free turbine is that the two turbines can operate at different speeds, and that these speeds can vary relative to each other. This is particularly advantageous for varying loads, such as turboprop engines.

Applications

Austin 250hp gas turbine, sectioned.

Most turboshaft and turboprop engines now use free turbines. This includes those for static power generation, as marine propulsion and particularly for helicopters.

Helicopters

Wessex HAS.31B showing the circumferential air intake of the Gazelle and the two exhausts (red covers) per side

A major market for turboshaft engines is that for helicopters. When turboshaft engines became available in the 1950s, they were rapidly adopted for both new designs and as replacements for piston engines. They offered more power and far better power to weight ratios. Helicopters of this period had barely adequate performance; the switch to a turbine engine could both reduce several hundred pounds of engine weight, 600 lb (270 kg) for the Napier Gazelle of the Westland Wessex, and also allow considerably more payload weight. For the Westland Whirlwind, this converted the inadequate piston-engined HAS.7 to the de Havilland Gnome turbine-powered HAR.9. As one of the first anti-submarine helicopters, the HAS.7 had been so restricted for weight that it could carry either a search sonar *or* an attack torpedo, but not both.

Ukrainian MS-14VM helicopter engine, with typical side-mounted exhaust and with the output power shaft from the turbine passing through it

The free-turbine engine was particularly favoured. It did not require a clutch, as the gas generator could be spun up to operating speed without requiring the output shaft to rotate. For the Wessex this was used to give a particularly fast take-off from a cold start. By locking the main rotor (and the power turbine) with the rotor brake, the engine could be spun up to operating speed, then lit, and when the engine core is at the operating speed of 10,500 rpm the brake is released and drive to the rotor smoothly increased as the power turbine gains speed. This was used to bring the rotor to speed from stationary in just 15 seconds and a time from engine start to take-off of only 30 seconds.

A further advantage of the free turbine design was the ease with which a counter-rotating engine could be designed and manufactured, simply by reversing the power turbine alone. This allowed handed engines to be made in pairs, when needed. It also allowed contra-rotating engines, where gas generator core and power turbine revolved in opposite directions, reducing the overall moment of inertia. For the helicopter engine replacement market, this ability allowed previous engines of either direction to be replaced simply. Some turboshaft engine's omni-angle freedom of their installation angle also allowed installation into existing helicopter designs, no matter how the previous engines had been arranged. In time though, the move towards axial LP compressors and so smaller diameter engines encouraged a move to the now standard layout of one or two engines set side-by-side, horizontally above the cabin.

Pusher Propfans

An attractively simple configuration making use of the free turbine is the propfan engine, with a rear-mounted unducted fan in pusher configuration, rather than the more familiar tractor layout. The first such engine was the very early and promising Metropolitan-Vickers F.3 of 1942 with a ducted fan, followed by the unducted and much lighter F.5. Development of these engines stopped abruptly owing to corporate takeovers, rather than technical reasons. Rolls-Royce continued with design studies for such engines into the 1980s, as did GE, but they have yet to appear as commercial engines.

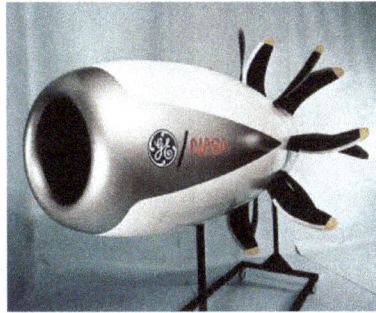

GE36 unducted fan

The advantage of the pusher propfan with a free power turbine is its simplicity. The prop blades are attached directly to the outside of the rotating turbine disc. No gearboxes or drive shafts are required. The short length of the rotating components also reduces vibration. The static structure of the engine over this length is a large diameter tube within the turbine. In most designs, two contra-rotating rings of turbine and propeller are used. Intermeshed contra-rotating turbines can act as the guide vanes for each other, removing the need for static vanes.

Risk of overspeed

A drawback to the simple free turbine turboprop is its behaviour if the load suddenly falls to zero. In such a case, the unconstrained free turbine overspeeds and will be destroyed by centrifugal forces.

Such a failure was the cause of the second prototype Bristol Britannia, G-ALRX's 1954 accident in the Severn Estuary. A failure in the propeller reduction gearbox led to an overspeed and destruction of the power turbine of N°3 engine. In the confined space of the Britannia's Bristol Proteus engine, fragments perforated the oil tank and led to a fire, which threatened the integrity of the wing spar. The pilot, Bill Pegg, then made a forced landing on the estuary mud.

To avoid such accidents, free turbine engines, including the Proteus, are now commonly fitted with a device to shut off the fuel supply at the HP cock if torque in the turbine output shaft suddenly falls to zero.

Radial turbine

A radial turbine is a turbine in which the flow of the working fluid is radial to the shaft. The difference between axial and radial turbines consists in the way the fluid flows through the components (compressor and turbine). Whereas for an axial turbine the rotor is 'impacted' by the fluid flow, for a radial turbine, the flow is smoothly orientated perpendicular to the rotation axis, and it drives the turbine in the same way water drives a watermill. The result is less mechanical stress (and less thermal stress, in case of hot working fluids) which enables a radial turbine to be simpler, more robust, and more efficient (in a similar power range) when compared to axial turbines. When it comes to high power ranges (above 5 MW) the radial turbine is no longer competitive (due to heavy and expensive rotor) and the efficiency becomes similar to that of the axial turbines.

Radial Turbine

Advantages and Challenges

Compared to an axial flow turbine, a radial turbine can employ a relatively higher pressure ratio (≈ 4) per stage with lower flow rates. Thus these machines fall in the lower specific speed and power ranges. For high temperature applications rotor blade cooling in radial stages is not as easy as in axial turbine stages. Variable angle nozzle blades can give higher stage efficiencies in a radial turbine stage even at off-design point operation. In the family of hydro-turbines, Francis turbine is a very well-known IFR turbine which generates much larger power with a relatively large impeller.

Components of Radial Turbines

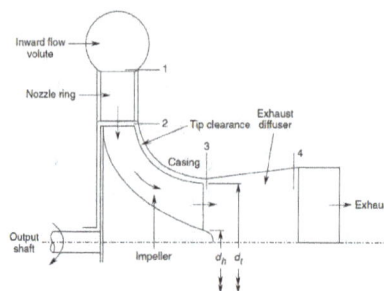

Ninety degree inward-flow radial turbine stage

Velocity triangles for an inward-flow radial (IFR) turbine stage with cantilever blades

The radial and tangential components of the absolute velocity c_2 are c_{r2} and c_{q2}, respectively. The relative velocity of the flow and the peripheral speed of the rotor are w_2 and u_2 respectively. The air angle at the rotor blade entry is given by

$$\tan \beta_2 = \frac{c_{r2}}{(c_{\theta 2} - u_2)}$$

Enthalpy and Entropy Diagram

The stagnation state of the gas at the nozzle entry is represented by point 01. The gas expands adiabatically in the nozzles from a pressure p_1 to p_2 with an increase in its velocity from c_1 to c_2. Since this is an energy transformation process, the stagnation enthalpy remains constant but the stagnation pressure decreases ($p_{01} > p_{02}$) due to losses. The energy transfer accompanied by an energy transformation process occurs in the rotor.

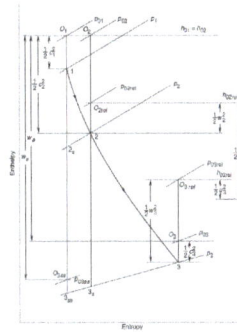

Enthalpy-entropy diagram for flow through an IFR turbine stage

Spouting Velocity

A reference velocity (c_o) known as the isentropic velocity, spouting velocity or stage terminal velocity is defined as that velocity which will be obtained during an isentropic expansion of the gas between the entry and exit pressures of the stage.

$$C_0 = \sqrt{2 C_p T_{01} \left(1 - \left(\frac{p_3}{p_{01}}\right)^{\frac{\gamma-1}{\gamma}}\right)}$$

Stage Efficiency

The total-to-static efficiency is based on this value of work.

$$\eta_t s = \frac{(h_{01} - h_{03})}{(h_{01} - h_{3ss})}$$

$$\eta_t s = \frac{\psi u_2^2}{\left[C_p T_{01}\left(1 - \left(\frac{p_3}{p_{01}}\right)^{\frac{(\gamma-1)}{\gamma}}\right)\right]}$$

Degree of Reaction

The relative pressure or enthalpy drop in the nozzle and rotor blades are determined by the degree of reaction of the stage. This is defined by

R= *static enthaply drop in rotor" stagnation enthalpy drop in stage"*

The two quantities within the parentheses in the numerator may have the same or opposite signs. This, besides other factors, would also govern the value of reaction. The stage reaction decreases as $C_{\theta 2}$ increases because this results in a large proportion of the stage enthalpy drop to occur in the nozzle ring.

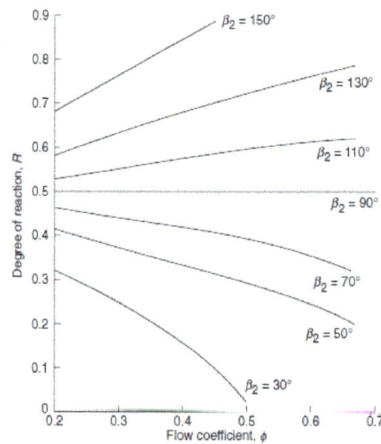

Variation of the degree of reaction with flow coefficient and air angle at rotor entry

Stage Losses

The stage work is less than the isentropic stage enthalpy drop on account of aerodynamic losses in the stage. The actual output at the turbine shaft is equal to the stage work minus the losses due to rotor disc and bearing friction.

- (a) skin friction and separation losses in the scroll and nozzle ring

They depend on the geometry and the coefficient of skin friction of these components.

- (b) Skin friction and separation losses in the rotor blade channels

These losses are also governed by the channel geometry, coefficient of skin friction and the ratio of the relative velocities w_3/w_2. In the ninety degree IFR turbine stage, the losses occurring in the radial and axial sections of the rotor are sometimes separately considered.

- (c) Skin friction and separation losses in the diffuser

These are mainly governed by the geometry of the diffuser and the rate of diffusion.

- (d) Secondary losses

These are due to circulatory flows developing into the various flow passages and are principally governed by the aerodynamic loading of the blades. The main parameters governing these losses are b_2/d_2, d_3/d_2 and hub-tip ratio at the rotor exit.

- *(e) Shock or incidence losses*

At off-design operation, there are additional losses in the nozzle and rotor blade rings on account of incidence at the leading edges of the blades. This loss is conventionally referred to as shock loss though it has nothing to do with the shock waves.

- *(f) Tip clearance loss*

This is due to the flow over the rotor blade tips which does not contribute to the energy transfer.

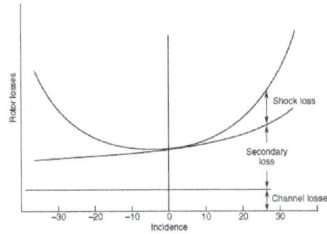

Losses in the rotor of an IFR turbine stage

Blade to Gas Speed Ratio

The blade-to-gas speed ratio can be expressed in terms of the isentropic stage terminal velocity c_0.

$$\sigma_s = \frac{u_2}{c_0} = [2(1 + \phi_2 \cot \beta_2)]^{\frac{-1}{2}}$$

for $\beta_2 = 90°$ $\sigma_s = 0.707$

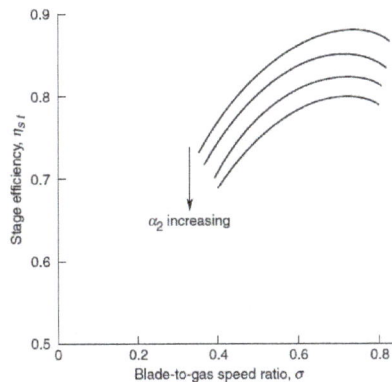

Variation of stage efficiency of an IFR turbine with blade-to-isentropic gas speed ratio

Outward-Flow Radial Stages

In outward flow radial turbine stages, the flow of the gas or steam occurs from smaller to larger diameters. The stage consists of a pair of fixed and moving blades. The increasing area of cross-section at larger diameters accommodates the expanding gas.

This configuration did not become popular with the steam and gas turbines. The only one which is employed more commonly is the Ljungstrom double rotation type turbine. It consists of rings of cantilever blades projecting from two discs rotating in opposite directions. The relative peripheral velocity of blades in two adjacent rows, with respect to each other, is high. This gives a higher value of enthalpy drop per stage.

Nikola Tesla's Bladeless Radial Turbine

In the early 1900s, Nikola Tesla developed and patented his Bladeless Turbine. One of the difficulties with bladed turbines is the complex and highly precise requirements for balancing and manufacturing the bladed rotor which has to be very well balanced. The blades are subject to corrosion and cavitation. Tesla attacked this problem by substituting a series of closely spaced disks for the blades of the rotor. The working fluid flows between the disks and transfers its energy to the rotor by means of the boundary layer effect or adhesion and viscosity rather than by impulse or reaction. Tesla stated his turbine could realize incredibly high efficiencies by steam. There has been no documented evidence yet of Tesla turbines achieving the efficiencies Tesla claimed.

In 2003 Scott O'Hearen took a patent on the Radial turbine blade system. This invention utilizes a combination of the concepts of a smooth runner surface for working fluid frictional contact and that of blades projecting axially from plural transverse runner faces. Author, Harikishan Gupta E., & Author, Shyam P. Kodali (2013). Design and Operation of Tesla Turbo machine - A state of the art review. International Journal of Advanced Transport Phenomena, 2(1), 2-3.

References

- Heiser, William H.; Pratt, David T. (1994). Hypersonic Airbreathing Propulsion. AIAA Education Series. Washington, D.C.: American Institute of Aeronautics and Astronautics. pp. 23–4. ISBN 1-56347-035-7.

- "The Avro Type 698 Vulcan" David W. Fildes, Pen & Sword Aviation 2012, ISBN 978 1 84884 284 7, p.301, Gas Floow Diagram

- "747 Creating the world's first jumbo jet and other adventures from a life in aviation" Joe Sutter, Smithsonian Books, ISBN 978-0-06-088241-9.

- Gas turbine aero-thermodynamics : with special reference to aircraft propulsion Sir Frank Whittle, Pergamon Press Ltd. 1981, ISBN 9780080267197.

- Bharat Verma (January 1, 2013). Indian Defence Review: Apr-Jun 2012. Lancer Publishers. p. 18. ISBN 978-81-7062-259-8. Retrieved October 25, 2015.

- Frank Northen Magill, ed. (1993). Magill's Survey of Science: Applied science series, Volume 3. Salem Press. p. 1431. ISBN 9780893567088.

- Ordway, Frederick I, III; Sharpe, Mitchell R (1979). The Rocket Team. Apogee Books Space Series 36. New York: Thomas Y. Crowell. p. 140. ISBN 1-894959-00-0.

- Wolf, William (2005). Boeing B-29 Superfortress: the ultimate look : from drawing board to VJ-Day. Schiffer. p. 205. ISBN 0764322575.

- Livingstone, Bob (1998). Under the Southern Cross: The B-24 Liberator in the South Pacific. Turner Publishing Company. p. 162. ISBN 1563114321.

Concepts and Theories of Gas Turbines

The Brayton cycle, a thermodynamic cycle can be used to explain the workings of a heat engine. The chapter elucidates on the salient features of the Brayton cycle and its two types. The chapter also explores the concepts of an internal combustion engine to aid in the better understanding of gas turbines.

Brayton Cycle

The Brayton cycle is a thermodynamic cycle that describes the workings of a constant pressure heat engine. The original Brayton engines used piston-compressor and expander systems, but more modern gas turbine engines and airbreathing jet engines also follow the Brayton cycle. Although the cycle is usually run as an open system (and indeed *must* be run as such if internal combustion is used), it is conventionally assumed for the purposes of thermodynamic analysis that the exhaust gases are reused in the intake, enabling analysis as a closed system.

The engine cycle is named after George Brayton (1830–1892), the American engineer who developed it originally for use in piston engines , although it was originally proposed and patented by Englishman John Barber in 1791. It is also sometimes known as the Joule cycle. The Ericsson cycle is similar to the Brayton cycle but uses external heat and incorporates the use of a regenerator. There are two types of Brayton cycles, open to the atmosphere and using internal combustion chamber or closed and using a heat exchanger.

History

In 1872, George Brayton applied for a patent for his "Ready Motor," a reciprocating constant pressure engine. The engine used a separate piston compressor and expander, with compressed air heated by internal fire as it entered the expander cylinder. The first versions of the Brayton engine mixed vaporized fuel with air as it entered the compressor by means of a heated-surface carburetor., The fuel / air was contained in a reservoir / tank and then it was admitted to the expansion cylinder and burned. As the fuel / air mixture entered the expansion cylinder it was ignited by a pilot flame. A screen was used to prevent the fire from entering / returning to the reservoir. In early versions of the engine, this screen sometimes failed and an explosion would occur, but in 1874 Brayton solved the explosion problem by adding the fuel just prior to the expander cylinder. The engine now used heavier fuels such as kerosine and fuel oil. Ignition remained pilot flame. Brayton produced and sold "Ready Motors" to perform a variety of tasks like water pumping, mill operation, even marine propulsion. Critics of the day claimed the engines ran smoothly and had an efficiency of about 17%.

FIG. 40.—Brayton Petroleum Engine.

Brayton petroleum engine

A Brayton cycle piston engine made from historical documents

Brayton cycle engines were some of the first internal combustion engines used for motive power. In 1881, John Holland used a Brayton engine to power the world's first successful self-propelled submarine, the Fenian Ram. John Philip Holland's submarine is preserved in the Paterson Museum in the Old Great Falls Historic District of Paterson, New Jersey.

George B Selden driving a Brayton powered automobile in 1905

In 1878, George B. Selden produced the first internal combustion automobile. Inspired by the internal combustion engine invented by George Brayton displayed at the Centennial Exposition in Philadelphia in 1876, Selden began working on a smaller lighter version, succeeding by 1878, some eight years before the public introduction of the Benz Patent Motorwagen in Europe. The Selden auto was powered by a 3-cylinder, 400-pound version of the Brayton Cycle engine which featured an enclosed crankshaft. Selden designed and constructed the engine with the help of Rochester machinist Frank H. Clement and his assistant William Gomm. He filed for a patent on May 8, 1879 (in a historical cross of people, the witness Selden chose was a local bank-teller, George Eastman, later to become famous for the Kodak camera). His application included not only the engine but its use in a 4 wheeled car. He then filed a series of amendments to his application which stretched out the legal process resulting in a delay of 16 years before the patent was granted on November 5, 1895. Henry Ford fought the Selden patent. Ford argued his cars used the four-stroke Otto cycle and not the Brayton engine shown used in the Selden auto. Ford won the appeal of the original case.

Models

A Brayton-type engine consists of three components:

- a compressor
- a mixing chamber
- an expander

In the original 19th-century Brayton engine, ambient air is drawn into a piston compressor, where it is compressed; ideally an isentropic process. The compressed air then runs through a mixing chamber where fuel is added, an isobaric process. The pressurized air and fuel mixture is then ignited in an expansion cylinder and energy is released, causing the heated air and combustion products to expand through a piston/cylinder; another ideally isentropic process. Some of the work extracted by the piston/cylinder is used to drive the compressor through a crankshaft arrangement.

The term Brayton cycle has more recently been given to the gas turbine engine. This also has three components:

- a gas compressor
- a burner (or combustion chamber)
- an expansion turbine

Ideal Brayton cycle:

- isentropic process – ambient air is drawn into the compressor, where it is pressurized.
- isobaric process – the compressed air then runs through a combustion chamber, where fuel is burned, heating that air—a constant-pressure process, since the chamber is open to flow in and out.
- isentropic process – the heated, pressurized air then gives up its energy, expanding through a turbine (or series of turbines). Some of the work extracted by the turbine is used to drive the compressor.
- isobaric process – heat rejection (in the atmosphere).

Actual Brayton cycle:

- adiabatic process – compression

- isobaric process – heat addition

- adiabatic process – expansion

- isobaric process – heat rejection

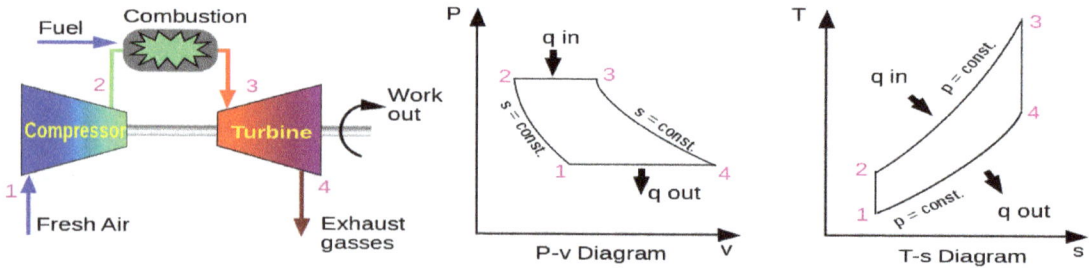

Idealized Brayton cycle

Since neither the compression nor the expansion can be truly isentropic, losses through the compressor and the expander represent sources of inescapable working inefficiencics. In general, increasing the compression ratio is the most direct way to increase the overall power output of a Brayton system.

The efficiency of the ideal Brayton cycle is $\eta = 1 - \dfrac{T_1}{T_2} = 1 - \left(\dfrac{P_1}{P_2}\right)^{(\gamma-1)/\gamma}$, where γ is the heat capacity ratio. Figure 1 indicates how the cycle efficiency changes with an increase in pressure ratio. Figure 2 indicates how the specific power output changes with an increase in the gas turbine inlet temperature for two different pressure ratio values.

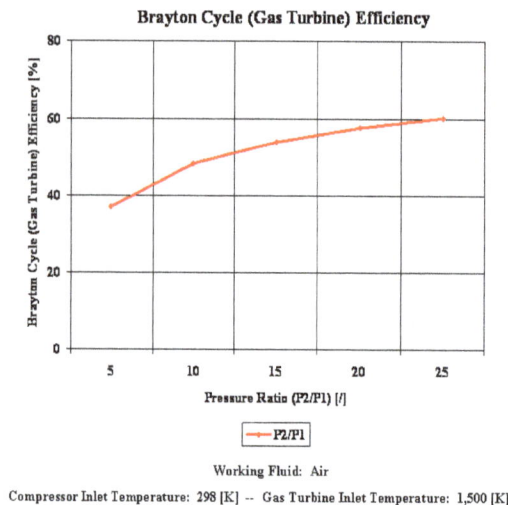

Working Fluid: Air

Compressor Inlet Temperature: 298 [K] -- Gas Turbine Inlet Temperature: 1,500 [K]

Figure1: Brayton cycle efficiency

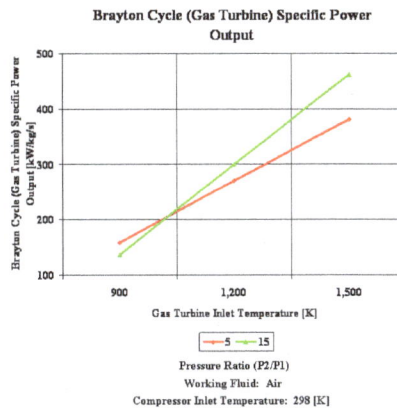

Figure2: Brayton cycle specific power output

The highest temperature in the cycle occurs at the end of the combustion process, and it is limited by the maximum temperature that the turbine blades can withstand. This also limits the pressure ratios that can be used in the cycle. For a fixed turbine inlet temperature, the net work output per cycle increases with the pressure ratio (thus the thermal efficiency) and the net work output. With less work output per cycle, a larger mass flow rate (thus a larger system) is needed to maintain the same power output, which may not be economical. In most common designs, the pressure ratio of a gas turbine ranges from about 11 to 16.

Methods to Increase Power

The power output of a Brayton engine can be improved in the following manners:

Reheat, wherein the working fluid—in most cases air—expands through a series of turbines, then is passed through a second combustion chamber before expanding to ambient pressure through a final set of turbines. This has the advantage of increasing the power output possible for a given compression ratio without exceeding any metallurgical constraints (typically about 1000 °C). The use of an afterburner for jet aircraft engines can also be referred to as "reheat"; it is a different process in that the reheated air is expanded through a thrust nozzle rather than a turbine. The metallurgical constraints are somewhat alleviated, enabling much higher reheat temperatures (about 2000 °C). Reheat is most often used to improve the specific power (per throughput of air), and is usually associated with a drop in efficiency, this effect is especially pronounced in afterburners due to the extreme amounts of extra fuel used.

Overspray, wherein, after a first compressor stage, water is injected into the compressor, thus increasing the mass-flow inside the compressor, increasing the turbine output power significantly and reducing compressor outlet temperatures. In a second compressor stage the water is completely converted to a gas form, offering some intercooling via its latent heat of vaporization.

Methods to Improve Efficiency

The efficiency of a Brayton engine can be improved in the following manners:

Increasing pressure ratio – As Figure 1 above shows, increasing the pressure ratio increases the efficiency of the Brayton cycle. This is analogous to the increase of efficiency seen in the Otto cycle when the compression ratio is increased. However, there are practical limits when it comes to increasing the pressure ratio. First of all, increasing the pressure ratio increases the compressor discharge temperature. This can cause the temperature of the gasses leaving the combustor to exceed the metallurgical limits of the turbine. Also, the diameter of the compressor blades becomes progressively smaller in higher pressure stages of the compressor. Because the gap between the blades and the engine casing increases in size as a percentage of the compressor blade height as the blades get smaller in diameter, a greater percentage of the compressed air can leak back past the blades in higher pressure stages. This causes a drop in compressor efficiency, and is most likely to occur in smaller gas turbines (since blades are inherently smaller to begin with). Finally, as can be seen in Figure 1, the efficiency levels off as pressure ratio increases. Hence, there is little to gain by increasing the pressure ratio further if it is already at a high level.

Recuperator – If the Brayton cycle is run at a low pressure ratio and a high temperature increase in the combustion chamber, the exhaust gas (after the last turbine stage) might still be hotter than the compressed inlet gas (after the last compression stage but before the combustor). In that case, a heat exchanger can be used to transfer thermal energy from the exhaust to the already compressed gas, before it enters the combustion chamber. The thermal energy transferred is effectively re-used, thus increasing efficiency. However, this form of heat recycling is only possible, if the engine is run in a low efficiency mode with low pressure ratio in the first place. Note, that transferring heat from the outlet (after the last turbine) to the inlet (before the first compressor stage) would reduce efficiency, as hotter inlet air means more volume and thus more work for the compressor. For engines with liquid cryogenic fuels, namely Hydrogen, it might be feasible, though, to use the fuel to cool the inlet air before compression to increase efficiency. This concept is extensively studied for the SABRE engine.

A Brayton engine also forms half of the 'combined cycle' system, which combines with a Rankine engine to further increase overall efficiency. However, although this increases overall efficiency, it does not actually increase the efficiency of the Brayton cycle itself.

Cogeneration systems make use of the waste heat from Brayton engines, typically for hot water production or space heating.

Variants

Closed Brayton Cycle

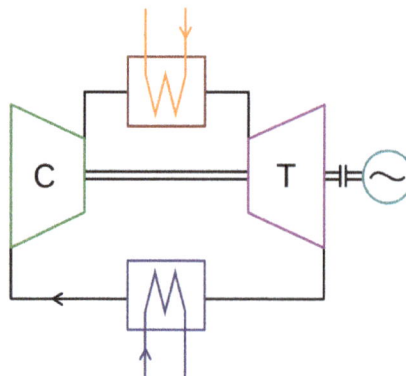

Closed Brayton cycle

C compressor and T turbine assembly

w high-temperature heat exchanger

м low-temperature heat exchanger

~ mechanical load, e.g. electric generator

A closed Brayton cycle recirculates the working fluid, the air expelled from the turbine is reintroduced into the compressor, this cycle uses a heat exchanger to heat the working fluid instead of an internal combustion chamber. The closed Brayton cycle is used for example in closed-cycle gas turbine and space power generation.

Solar Brayton Cycle

In 2002, a hybrid open solar Brayton cycle was operated for the first time consistently and effectively with relevant papers published, in the frame of the EU SOLGATE program. The air was heated from 570 K to over 1000 K into the combustor chamber. Further hybridization was achieved during the EU Solhyco project running a hybridized Brayton cycle with solar energy and Biodiesel only. This technology was scaled up to 4.6 MW within the project Solugas located near Seville where it is currently demonstrated at pre-commercial scale.

Reverse Brayton Cycle

A Brayton cycle that is driven in reverse, via net work input, and when air is the working fluid, is the air refrigeration cycle or Bell Coleman cycle. Its purpose is to move heat, rather than produce work. This air cooling technique is used widely in jet aircraft for air conditioning systems utilizing air tapped from the engine compressors.

Internal Combustion Engine

"ICEV" redirects here. For the form of water ice, see Ice V. For the high speed train, see ICE V.

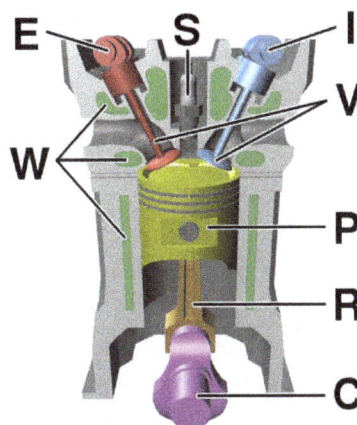

Diagram of a cylinder as found in 4-stroke gasoline engines.:

C – crankshaft.
E – exhaust camshaft.
I – inlet camshaft.
P – piston.
R – connecting rod.
S – spark plug.
V – valves. red: exhaust, blue: intake.
W – cooling water jacket.
gray structure – engine block.

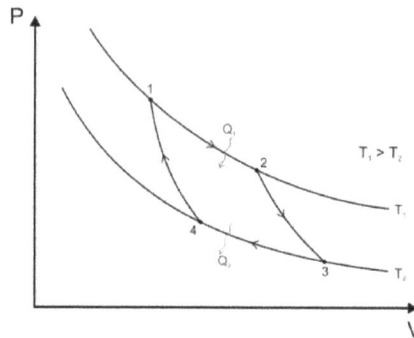

Diagram describing the ideal combustion cycle by Carnot

An internal combustion engine (ICE) is a heat engine where the combustion of a fuel occurs with an oxidizer (usually air) in a combustion chamber that is an integral part of the working fluid flow circuit. In an internal combustion engine the expansion of the high-temperature and high-pressure gases produced by combustion apply direct force to some component of the engine. The force is applied typically to pistons, turbine blades, rotor or a nozzle. This force moves the component over a distance, transforming chemical energy into useful mechanical energy.

The first commercially successful internal combustion engine was created by É tienne L enoir around 1859 and the first modern internal combustion engine was created in 1876 by Nikolaus Otto.

The term *internal combustion engine* usually refers to an engine in which combustion is intermittent, such as the more familiar four-stroke and two-stroke piston engines, along with variants, such as the six-stroke piston engine and the Wankel rotary engine. A second class of internal combustion engines use continuous combustion: gas turbines, jet engines and most rocket engines, each of which are internal combustion engines on the same principle as previously described. Firearms are also a form of internal combustion engine.

Internal combustion engines are quite different from external combustion engines, such as steam or Stirling engines, in which the energy is delivered to a working fluid not consisting of, mixed with, or contaminated by combustion products. Working fluids can be air, hot water, pressurized water or even liquid sodium, heated in a boiler. ICEs are usually powered by energy-dense fuels such as gasoline or diesel, liquids derived from fossil fuels. While there are many stationary applications, most ICEs are used in mobile applications and are the dominant power supply for vehicles such as cars, aircraft, and boats.

Typically an ICE is fed with fossil fuels like natural gas or petroleum products such as gasoline, diesel fuel or fuel oil. There's a growing usage of renewable fuels like biodiesel for compression ignition engines and bioethanol or methanol for spark ignition engines. Hydrogen is sometimes used, and can be made from either fossil fuels or renewable energy.

History

Various scientists and engineers contributed to the development of internal combustion engines. In 1791, John Barber developed a turbine. In 1798, John Stevens built the first internal combustion engine. In 1807, Swiss engineer François Isaac de Rivaz built an internal combustion engine ignited by electric spark. In 1823, Samuel Brown patented the first internal combustion engine to be applied industrially.

In 1860, Belgian Jean Joseph Etienne Lenoir produced a gas-fired internal combustion engine. In 1864, Nikolaus Otto patented the first atmospheric gas engine. In 1872, American George Brayton invented the first commercial liquid-fuelled internal combustion engine. In 1876, Nikolaus Otto, working with Gottlieb Daimler and Wilhelm Maybach, patented the compressed charge, four-cycle engine. In 1879, Karl Benz patented a reliable two-stroke gas engine. In 1892, Rudolf Diesel developed the first compressed charge, compression ignition engine. In 1926, Robert Goddard launched the first liquid-fueled rocket. In 1939, the Heinkel He 178 became the world's first jet aircraft.

Etymology

At one time, the word *engine* (via Old French, from Latin *ingenium*, "ability") meant any piece of machinery — a sense that persists in expressions such as *siege engine*. A "motor" (from Latin *motor*, "mover") is any machine that produces mechanical power. Traditionally, electric motors are not referred to as "Engines"; however, combustion engines are often referred to as "motors." (An *electric engine* refers to a locomotive operated by electricity.)

In boating an internal combustion engine that is installed in the hull is referred to as an engine, but the engines that sit on the transom are referred to as motors.

Applications

Reciprocating engine as found inside a car

Reciprocating piston engines are by far the most common power source for land and water vehicles, including automobiles, motorcycles, ships and to a lesser extent, locomotives (some are electrical but most use Diesel engines). Rotary engines of the Wankel design are used in some automobiles, aircraft and motorcycles.

Where very high power-to-weight ratios are required, internal combustion engines appear in the form of combustion turbines or Wankel engines. Powered aircraft typically uses an ICE which may be a reciprocating engine. Airplanes can instead use jet engines and helicopters can instead employ turboshafts; both of which are types of turbines. In addition to providing propulsion, airliners may employ a separate ICE as an auxiliary power unit. Wankel engines are fitted to many unmanned aerial vehicles.

Big Diesel generator used for backup power

ICEs drive some of the large electric generators that power electrical grids. They are found in the form of combustion turbines in combined cycle power plants with a typical electrical output in the range of 100 MW to 1 GW. The high temperature exhaust is used to boil and superheat water to run a steam turbine. Thus, the efficiency is higher because more energy is extracted from the fuel than what could be extracted by the combustion turbine alone. In combined cycle power plants efficiencies in the range of 50% to 60% are typical. In a smaller scale Diesel generators are used for backup power and for providing electrical power to areas not connected to an electric grid.

Combined cycle power plant

Small engines (usually 2-stroke gasoline engines) are a common power source for lawnmowers, string trimmers, chain saws, leafblowers, pressure washers, snowmobiles, jet skis, outboard motors, mopeds, and motorcycles.

Classification

There are several possible ways to classify internal combustion engines.

Reciprocating:

By number of strokes

- Two-stroke engine
- Clerk Cycle 1879
- Day Cycle
- Four-stroke engine (Otto cycle)
- Six-stroke engine

By type of ignition

- Compression-ignition engine
- Spark-ignition engine (commonly found as gasoline engines)

By mechanical/thermodynamical cycle (these 2 cycles do not encompass all reciprocating engines, and are infrequently used):

- Atkinson cycle
- Miller cycle

Rotary:

- Wankel engine

Continuous Combustion:

- Gas turbine
- Jet engine
  - Rocket engine
  - Ramjet

  The following jet engine types are also gas turbines types:

  - Turbojet
  - Turbofan
  - Turboprop

Reciprocating Engines

Structure

The base of a reciprocating internal combustion engine is the engine block, which is typically made of cast iron or aluminium. The engine block contains the cylinders. In engines with more than one cylinder they are usually arranged either in 1 row (straight engine) or 2 rows (boxer engine or V engine); 3 rows are occasionally used (W engine) in contemporary engines, and other engine configurations are possible and have been used. Single cylinder engines are common for motorcycles and in small engines of machinery. Water-cooled engines contain passages in the engine block where cooling fluid circulates (the water jacket). Some small engines are air-cooled, and instead of having a water jacket the cylinder block has fins protruding away from it to cool by directly transferring heat to the air. The cylinder walls are usually finished by honing to obtain a cross hatch, which is better able to retain the oil. A too rough surface would quickly harm the engine by excessive wear on the piston.

Bare cylinder block of a V8 engine

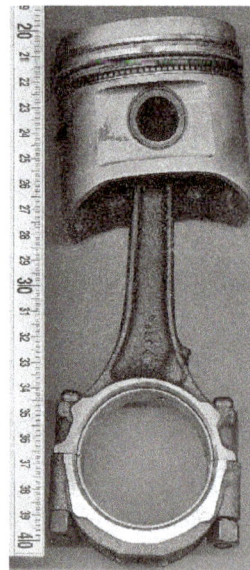

Piston, piston ring, gudgeon pin and connecting rod

The pistons are short cylindrical parts which seal one end of the cylinder from the high pressure of the compressed air and combustion products and slide continuously within it while the engine is in operation. The top wall of the piston is termed its *crown* and is typically flat or concave. Some two-stroke engines use pistons with a deflector head. Pistons are open at the bottom and hollow except for an integral reinforcement structure (the piston web). When an engine is working the gas pressure in the combustion chamber exerts a force on the piston crown which is transferred through its web to a gudgeon pin. Each piston has rings fitted around its circumference that mostly prevent the gases from leaking into the crankcase or the oil into the combustion chamber. A ventilation system drives the small amount of gas that escape past the pistons during normal operation (the blow-by gases) out of the crankcase so that it does not accumulate contaminating the oil and creating corrosion. In two-stroke gasoline engines the crankcase is part of the air–fuel path and due to the continuous flow of it they do not need a separate crankcase ventilation system.

Valve train above a Diesel engine cylinder head. This engine uses rocker arms but no pushrods.

The cylinder head is attached to the engine block by numerous bolts or studs. It has several functions. The cylinder head seals the cylinders on the side opposite to the pistons; it contains short ducts (the *ports*) for intake and exhaust and the associated intake valves that open to let the cylinder be filled with fresh air and exhaust valves that open to allow the combustion gases to escape. However, 2-stroke crankcase scavenged engines connect the gas ports directly to the cylinder wall without poppet valves; the piston controls their opening and occlusion instead. The cylinder head also holds the spark plug in the case of spark ignition engines and the injector for engines that use direct injection. All CI engines use fuel injection, usually direct injection but some engines instead use indirect injection. SI engines can use a carburetor or fuel injection as port injection or direct injection. Most SI engines have a single spark plug per cylinder but some have 2. A head gasket prevents the gas from leaking between the cylinder head and the engine block. The opening and closing of the valves is controlled by one or several camshafts and springs—or in some engines—a desmodromic mechanism that uses no springs. The camshaft may press directly the stem of the valve or may act upon a rocker arm, again, either directly or through a pushrod.

Engine block seen from below. The cylinders, oil spray nozzle and half of the main bearings are clearly visible.

The crankcase is sealed at the bottom with a sump that collects the falling oil during normal operation to be cycled again. The cavity created between the cylinder block and the sump houses a crankshaft that converts the reciprocating motion of the pistons to rotational motion. The crankshaft is held in place relative to the engine block by main bearings, which allow it to rotate.

Bulkheads in the crankcase form a half of every main bearing; the other half is a detachable cap. In some cases a single *main bearing deck* is used rather than several smaller caps. A connecting rod is connected to offset sections of the crankshaft (the crankpins) in one end and to the piston in the other end through the gudgeon pin and thus transfers the force and translates the reciprocating motion of the pistons to the circular motion of the crankshaft. The end of the connecting rod attached to the gudgeon pin is called its small end, and the other end, where it is connected to the crankshaft, the big end. The big end has a detachable half to allow assembly around the crankshaft. It is kept together to the connecting rod by removable bolts.

The cylinder head has an intake manifold and an exhaust manifold attached to the corresponding ports. The intake manifold connects to the air filter directly, or to a carburetor when one is present, which is then connected to the air filter. It distributes the air incoming from these devices to the individual cylinders. The exhaust manifold is the first component in the exhaust system. It collects the exhaust gases from the cylinders and drives it to the following component in the path. The exhaust system of an ICE may also include a catalytic converter and muffler. The final section in the path of the exhaust gases is the tailpipe.

4-Stroke Engines

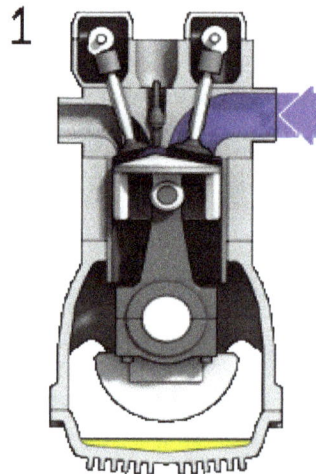

Diagram showing the operation of a 4-stroke SI engine. Labels:

1 - Induction
2 - Compression
3 - Power
4 - Exhaust

The *top dead center* (TDC) of a piston is the position where it is nearest to the valves; *bottom dead center* (BDC) is the opposite position where it is furthest from them. A *stroke* is the movement of a piston from TDC to BDC or vice versa together with the associated process. While an engine is in operation the crankshaft rotates continuously at a nearly constant speed. In a 4-stroke ICE each piston experiences 2 strokes per crankshaft revolution in the following order. Starting the description at TDC, these are:

1. Intake, induction or suction: The intake valves are open as a result of the cam lobe pressing

down on the valve stem. The piston moves downward increasing the volume of the combustion chamber and allowing air to enter in the case of a CI engine or an air fuel mix in the case of SI engines that do not use direct injection. The air or air-fuel mixture is called the *charge* in any case.

2. Compression: In this stroke, both valves are closed and the piston moves upward reducing the combustion chamber volume which reaches its minimum when the piston is at TDC. The piston performs work on the charge as it is being compressed; as a result its pressure, temperature and density increase; an approximation to this behavior is provided by the ideal gas law. Just before the piston reaches TDC, ignition begins. In the case of a SI engine, the spark plug receives a high voltage pulse that generates the spark which gives it its name and ignites the charge. In the case of a CI engine the fuel injector quickly injects fuel into the combustion chamber as a spray; the fuel ignites due to the high temperature.

3. Power or working stroke: The pressure of the combustion gases pushes the piston downward, generating more work than it required to compress the charge. Complementary to the compression stroke, the combustion gases expand and as a result their temperature, pressure and density decreases. When the piston is near to BDC the exhaust valve opens. The combustion gases expand irreversibly due to the leftover pressure—in excess of back pressure, the gauge pressure on the exhaust port—; this is called the *blowdown*.

4. Exhaust: The exhaust valve remains open while the piston moves upward expelling the combustion gases. For naturally aspirated engines a small part of the combustion gases may remain in the cylinder during normal operation because the piston does not close the combustion chamber completely; these gases dissolve in the next charge. At the end of this stroke, the exhaust valve closes, the intake valve opens, and the sequence repeats in the next cycle. The intake valve may open before the exhaust valve closes to allow better scavenging.

2-Stroke Engines

The defining characteristic of this kind of engine is that each piston completes a cycle every crankshaft revolution. The 4 processes of intake, compression, power and exhaust take place in only 2 strokes so that it is not possible to dedicate a stroke exclusively for each of them. Starting at TDC the cycle consist of:

1. Power: While the piston is descending the combustion gases perform work on it—as in a 4-stroke engine—. The same thermodynamic considerations about the expansion apply.

2. Scavenging: Around 75° of crankshaft rotation before BDC the exhaust valve or port opens, and blowdown occurs. Shortly thereafter the intake valve or transfer port opens. The incoming charge displaces the remaining combustion gases to the exhaust system and a part of the charge may enter the exhaust system as well. The piston reaches BDC and reverses direction. After the piston has traveled a short distance upwards into the cylinder the exhaust valve or port closes; shortly the intake valve or transfer port closes as well.

3. Compression: With both intake and exhaust closed the piston continues moving upwards compressing the charge and performing a work on it. As in the case of a 4-stroke engine,

ignition starts just before the piston reaches TDC and the same consideration on the thermodynamics of the compression on the charge.

While a 4-stroke engine uses the piston as a positive displacement pump to accomplish scavenging taking 2 of the 4 strokes, a 2-stroke engine uses the last part of the power stroke and the first part of the compression stroke for combined intake and exhaust. The work required to displace the charge and exhaust gases comes from either the crankcase or a separate blower. For scavenging, expulsion of burned gas and entry of fresh mix, two main approaches are described: Loop scavenging, and Uniflow scavenging, SAE news published in the 2010s that 'Loop Scavenging' is better under any circumstance than Uniflow Scavenging.

Crankcase Scavenged

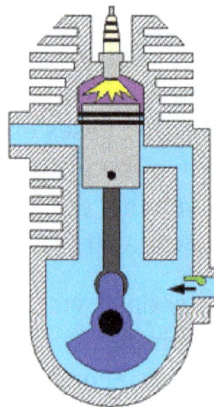

Diagram of a crankcase scavenged 2-stroke engine in operation

Some SI engines are crankcase scavenged and do not use poppet valves. Instead the crankcase and the part of the cylinder below the piston is used as a pump. The intake port is connected to the crankcase through a reed valve or a rotary disk valve driven by the engine. For each cylinder a transfer port connects in one end to the crankcase and in the other end to the cylinder wall. The exhaust port is connected directly to the cylinder wall. The transfer and exhaust port are opened and closed by the piston. The reed valve opens when the crankcase pressure is slightly below intake pressure, to let it be filled with a new charge; this happens when the piston is moving upwards. When the piston is moving downwards the pressure in the crankcase increases and the reed valve closes promptly, then the charge in the crankcase is compressed. When the piston is moving upwards, it uncovers the exhaust port and the transfer port and the higher pressure of the charge in the crankcase makes it enter the cylinder through the transfer port, blowing the exhaust gases. Lubrication is accomplished by adding *2-stroke oil* to the fuel in small ratios. *Petroil* refers to the mix of gasoline with the aforesaid oil. This kind of 2-stroke engines has a lower efficiency than comparable 4-strokes engines and release a more polluting exhaust gases for the following conditions:

They use a *total-loss lubrication system*: all the lubricating oil is eventually burned along with the fuel.

There are conflicting requirements for scavenging: On one side, enough fresh charge needs to be introduced in each cycle to displace almost all the combustion gases but introducing too much of it means that a part of it gets in the exhaust.

They must use the transfer port(s) as a carefully designed and placed nozzle so that a gas current is created in a way that it sweeps the whole cylinder before reaching the exhaust port so as to expel the combustion gases, but minimize the amount of charge exhausted. 4-stroke engines have the benefit of forcibly expelling almost all of the combustion gases because during exhaust the combustion chamber is reduced to its minimum volume. In crankcase scavenged 2-stroke engines, exhaust and intake are performed mostly simultaneously and with the combustion chamber at its maximum volume.

The main advantage of 2-stroke engines of this type is mechanical simplicity and a higher power-to-weight ratio than their 4-stroke counterparts. Despite having twice as many power strokes per cycle, less than twice the power of a comparable 4-stroke engine is attainable in practice.

In the USA two stroke motorcycle and automobile engines were banned due to the pollution, although many thousands of lawn maintenance engines are in use.

Blower Scavenged

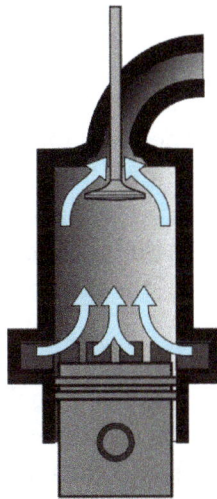

Diagram of uniflow scavenging

Using a separate blower avoids many of the shortcomings of crankcase scavenging, at the expense of increased complexity which means a higher cost and an increase in maintenance requirement. An engine of this type uses ports or valves for intake and valves for exhaust, except opposed piston engines, which may also use ports for exhaust. The blower is usually of the Roots-type but other types have been used too. This design is commonplace in CI engines, and has been occasionally used in SI engines.

CI engines that use a blower typically use *uniflow scavenging*. In this design the cylinder wall contains several intake ports placed uniformly spaced along the circumference just above the position that the piston crown reaches when at BDC. An exhaust valve or several like that of 4-stroke engines is used. The final part of the intake manifold is an air sleeve which feeds the intake ports. The intake ports are placed at an horizontal angle to the cylinder wall (I.e: they are in plane of the piston crown) to give a swirl to the incoming charge to improve combustion. The largest reciprocating IC are low speed CI engines of this type; they are used for marine propulsion or electric power generation and achieve the highest thermal efficiencies among internal combustion engines of any kind. Some Diesel-electric locomotive engines operate on the 2-stroke cycle. The most powerful of them have a brake power of around 4.5 MW or 6,000 HP. The EMD SD90MAC class of locomotives use a 2-stroke engine. The comparable class GE AC6000CW whose prime mover has almost the same brake power uses a 4-stroke engine.

An example of this type of engine is the Wärtsilä-Sulzer RTA96-C turbocharged 2-stroke Diesel, used in large container ships. It is the most efficient and powerful internal combustion engine in the world with a thermal efficiency over 50%. For comparison, the most efficient small four-stroke engines are around 43% thermally-efficient (SAE 900648); size is an advantage for efficiency due to the increase in the ratio of volume to surface area.

Historical Design

Dugald Clerk developed the first two cycle engine in 1879. It used a separate cylinder which functioned as a pump in order to transfer the fuel mixture to the cylinder.

In 1899 John Day simplified Clerk's design into the type of 2 cycle engine that is very widely used today. Day cycle engines are crankcase scavenged and port timed. The crankcase and the part of the cylinder below the exhaust port is used as a pump. The operation of the Day cycle engine begins when the crankshaft is turned so that the piston moves from BDC upward (toward the head) creating a vacuum in the crankcase/cylinder area. The carburetor then feeds the fuel mixture into the crankcase through a reed valve or a rotary disk valve (driven by the engine). There are cast in ducts from the crankcase to the port in the cylinder to provide for intake and another from the exhausst port to the exhaust pipe. The height of the port in relationship to the length of the cylinder is called the "port timing."

On the first upstroke of the engine there would be no fuel inducted into the cylinder as the crankcase was empty. On the downstroke the piston now compresses the fuel mix, which has lubricated the piston in the cylinder and the bearings due to the fuel mix having oil added to it. As the piston moves downward is first uncovers the exhaust, but on the first stroke there is no burnt fuel to exhaust. As the piston moves downward further, it uncovers the intake port which has a duct that runs to the crankcase. Since the fuel mix in the crankcase is under pressure the mix moves through the duct and into the cylinder.

Because there is no obstruction in the cylinder of the fuel to move directly out of the exhaust port prior to the piston rising far enough to close the port, early engines used a high domed piston to slow down the flow of fuel. Later the fuel was "resonated" back into the cylinder using an expansion chamber design. When the piston rose close to TDC a spark ignites the fuel. As the piston is driven downward with power it first uncovers the exhaust port where the burned fuel is expelled under high pressure and then the intake port where the process has been completed and will keep repeating.

Later engines used a type of porting devised by the Deutz company to improve performance. It was called the Schnurle Reverse Flow system. DKW licensed this design for all their motorcycles. Their DKW RT 125 was one of the first motor vehicles to achieve over 100 mpg as a result.

Ignition

Internal combustion engines require ignition of the mixture, either by spark ignition (SI) or compression ignition (CI). Before the invention of reliable electrical methods, hot tube and flame methods were used. Experimental engines with laser ignition have been built.

Spark Ignition Process

Bosch Magneto

The spark ignition engine was a refinement of the early engines which used Hot Tube ignition. When Bosch developed the magneto it became the primary system for producing electricity to energize a spark plug. Many small engines still use magneto ignition. Small engines are started by hand cranking using a recoil starter or hand crank . Prior to Charles F. Kettering of Delco's development of the automotive starter all gasoline engined automobiles used a hand crank.

Larger engines typically power their starting motors and Ignition systems using using the electrical energy stored in a lead–acid battery. The battery's charged state is maintained by an automotive alternator or (previously) a generator which uses engine power to create electrical energy storage.

Points and Coil Ignition

The battery supplies electrical power for starting when the engine has a starting motor system, and supplies electrical power when the engine is off. The battery also supplies electrical power during rare run conditions where the alternator cannot maintain more than 13.8 volts (for a common 12V automotive electrical system). As alternator voltage falls below 13.8 volts, the lead-acid storage battery increasingly picks up electrical load. During virtually all running conditions, including normal idle conditions, the alternator supplies primary electrical power.

Some systems disable alternator field (rotor) power during wide open throttle conditions. Disabling the field reduces alternator pulley mechanical loading to nearly zero, maximizing crankshaft power. In this case the battery supplies all primary electrical power.

Gasoline engines take in a mixture of air and gasoline and compress it by the movement of the piston from bottom dead center to top dead center when the fuel is at maximum compression. The reduction in the size of the swept area of the cylinder and taking into account the volume of the combustion chamber is described by a ratio. Early engines had compression ratios of 6 to 1. As compression ratios were increased the efficiency of the engine increased as well.

With early induction and ignition systems the compression ratios had to be kept low. With advances in fuel technology and combustion management high performance engines can run reliably at 12:1 ratio. With low octane fuel a problem would occur as the compression ratio increased as the fuel was igniting due to the rise in temperature that resulted. Charles Kettering developed a lead additive which allowed higher compression ratios.

The fuel mixture is ignited at difference progressions of the piston in the cylinder. At low rpm the spark is timed to occur close to the piston achieving top dead center. In order to produce more power, as rpm rises the spark is advanced sooner during piston movement. The spark occurs while the fuel is still being compressed progressively more as rpm rises.

The necessary high voltage, typically 10,000 volts, is supplied by an induction coil or transformer. The induction coil is a fly-back system, using interruption of electrical primary system current through some type of synchronized interrupter. The interrupter can be either contact points or a power transistor. The problem with this type of ignition is that as RPM increases the available of electrical energy decreases. This is especially as problem since the amount of energy needed to ignite a more dense fuel mixture is higher. The result was often a high rpm misfire.

Capacitor discharge ignition was developed. It produces a rising voltage that is sent to the spark plug. CD system voltages can reach 60,000 volts. CD ignitions use step-up transformers. The step-up transformer uses energy stored in a capacitance to generate electric spark. With either system, a mechanical or electrical control system provides a carefully timed high-voltage to the proper cylinder. This spark, via the spark plug, ignites the air-fuel mixture in the engine's cylinders.

While gasoline internal combustion engines are much easier to start in cold weather than diesel engines, they can still have cold weather starting problems under extreme conditions. For years the solution was to park the car in heated areas. In some parts of the world the oil was actually drained and heated over night and returned to the engine for cold starts. In the early 1950s the gasoline Gasifier unit was developed, where, on cold weather starts, raw gasoline was diverted to the unit where part of the fuel was burned causing the other part to become a hot vapor sent directly to the intake valve manifold. This unit was quite popular until electric engine block heaters became standard on gasoline engines sold in cold climates.

Compression Ignition Process

Diesel, PPC (Partially premixed combustion) and HCCI (Homogeneous charge compression ignition) engines, rely solely on heat and pressure created by the engine in its compression process for ignition. The compression level that occurs is usually twice or more than a gasoline engine. Diesel engines take in air only, and shortly before peak compression, spray a small quantity of diesel fuel into the cylinder via a fuel injector that allows the fuel to instantly ignite. HCCI type engines take in both air and fuel, but continue to rely on an unaided auto-combustion process, due to higher pressures and heat. This is also why diesel and HCCI engines are more susceptible to cold-starting issues, although they run just as well in cold weather once started. Light duty diesel engines with indirect injection in automobiles and light trucks employ glowplugs (or other pre-heating: see Cummins IS-B#6BT) that pre-heat the combustion chamber just before starting to reduce no-start conditions in cold weather. Most diesels also have a battery and charging system; nevertheless, this system is secondary and is added by manufacturers as a luxury for the ease of starting, turning fuel on and off (which can also be done via a switch or mechanical apparatus), and for running auxiliary electrical components and accessories. Most new engines rely on electrical and electronic engine control units (ECU) that also adjust the combustion process to increase efficiency and reduce emissions.

Lubrication

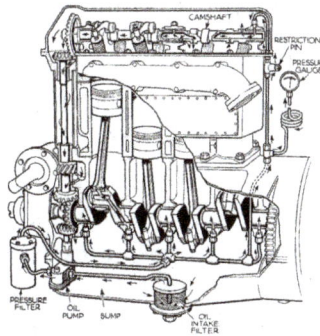

Diagram of an engine using pressurized lubrication

Surfaces in contact and relative motion to other surfaces require lubrication to reduce wear, noise and increase efficiency by reducing the power wasting in overcoming friction, or to make the mechanism work at all. At the very least, an engine requires lubrication in the following parts:

- Between pistons and cylinders

- Small bearings

- Big end bearings

- Main bearings

- Valve gear (The following elements may not be present):

    - Tappets

    - Rocker arms

    - Pushrods

    - Timing chain or gears. Toothed belts do not require lubrication.

In 2-stroke crankcase scavenged engines, the interior of the crankcase, and therefore the crankshaft, connecting rod and bottom of the pistons are sprayed by the 2-stroke oil in the air-fuel-oil mixture which is then burned along with the fuel. The valve train may be contained in a compartment flooded with lubricant so that no oil pump is required.

In a *splash lubrication system* no oil pump is used. Instead the crankshaft dips into the oil in the sump and due to its high speed, it splashes the crankshaft, connecting rods and bottom of the pistons. The connecting rod big end caps may have an attached scoop to enhance this effect. The valve train may also be sealed in a flooded compartment, or open to the crankshaft in a way that it receives splashed oil and allows it to drain back to the sump. Splash lubrication is common for small 4-stroke engines.

In a *forced* (also called *pressurized*) *lubrication system*, lubrication is accomplished in a closed loop which carries motor oil to the surfaces serviced by the system and then returns the oil to a reservoir. The auxiliary equipment of an engine is typically not serviced by this loop; for instance, an alternator may use ball bearings sealed with its lubricant. The reservoir for the oil is usually the sump, and when this is the case, it is called a *wet sump* system. When there is a different oil reservoir the crankcase still catches it, but it is continuously drained by a dedicated pump; this is called a *dry sump* system.

On its bottom, the sump contains an oil intake covered by a mesh filter which is connected to an oil pump then to an oil filter outside the crankcase, from there it is diverted to the crankshaft main bearings and valve train. The crankcase contains at least one *oil gallery* (a conduit inside a crankcase wall) to which oil is introduced from the oil filter. The main bearings contain a groove through all or half its circumference; the oil enters to these grooves from channels connected to the oil gallery. The crankshaft has drillings which take oil from these grooves and deliver it to the big end bearings. All big end bearings are lubricated this way. A single main bearing may provide oil for 0, 1 or 2 big end bearings. A similar system may be used to lubricate the piston, its gudgeon pin and the small end of its connecting rod; in this system, the connecting rod big end has a groove around the crankshaft and a drilling connected to the groove which distributes oil from there to the bottom of the piston and from then to the cylinder.

Other systems are also used to lubricate the cylinder and piston. The connecting rod may have a nozzle to throw an oil jet to the cylinder and bottom of the piston. That nozzle is in movement relative to the cylinder it lubricates, but always pointed towards it or the corresponding piston.

Typically a forced lubrication systems have a lubricant flow higher than what is required to lubricate satisfactorily, in order to assist with cooling. Specifically, the lubricant system helps to move heat from the hot engine parts to the cooling liquid (in water-cooled engines) or fins (in air-cooled engines) which then transfer it to the environment. The lubricant must be designed to be chemically stable and maintain suitable viscosities within the temperature range it encounters in the engine.

Cylinder Configuration

Common cylinder configurations include the straight or inline configuration, the more compact V configuration, and the wider but smoother flat or boxer configuration. Aircraft engines can also adopt a radial configuration, which allows more effective cooling. More unusual configurations such as the H, U, X, and W have also been used.

Multiple cylinder engines have their valve train and crankshaft configured so that pistons are at different parts of their cycle. It is desirable to have the piston's cycles uniformly spaced (this is called *even firing*) especially in forced induction engines; this reduces torque pulsations and makes inline engines with more than 3 cylinders statically balanced in its primary forces. However, some engine configurations require odd firing to achieve better balance than what is possible with even firing. For instance, a 4-stroke I2 engine has better balance when the angle between the crankpins is 180° because the pistons move in opposite directions and inertial forces partially cancel, but this gives an odd firing pattern where one cylinder fires 180° of crankshaft rotation after the other, then no cylinder fires for 540°. With an even firing pattern the pistons would move in unison and the associated forces would add.

Multiple crankshaft configurations do not necessarily need a cylinder head at all because they can instead have a piston at each end of the cylinder called an opposed piston design. Because fuel inlets and outlets are positioned at opposed ends of the cylinder, one can achieve uniflow scavenging, which, as in the four-stroke engine is efficient over a wide range of engine speeds. Thermal efficiency is improved because of a lack of cylinder heads. This design was used in the Junkers Jumo 205 diesel aircraft engine, using two crankshafts at either end of a single bank of cylinders, and most remarkably in the Napier Deltic diesel engines. These used three crankshafts to serve three banks of double-ended cylinders arranged in an equilateral triangle with the crankshafts at the corners. It was also used in single-bank locomotive engines, and is still used in marine propulsion engines and marine auxiliary generators.

Diesel Cycle

Most truck and automotive diesel engines use a cycle reminiscent of a four-stroke cycle, but with a compression heating ignition system, rather than needing a separate ignition system. This variation is called the diesel cycle. In the diesel cycle, diesel fuel is injected directly into the cylinder so that combustion occurs at constant pressure, as the piston moves.

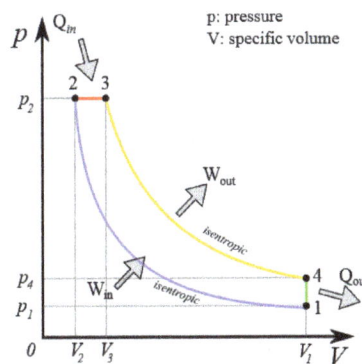

P-v Diagram for the Ideal Diesel cycle. The cycle follows the numbers 1–4 in clockwise direction.

Otto cycle: Otto cycle is the typical cycle for most of the cars internal combustion engines, that work using gasoline as a fuel. Otto cycle is exactly the same one that was described for the four-stroke engine. It consists of the same four major steps: Intake, compression, ignition and exhaust.

PV diagram for Otto cycle On the PV-diagram, 1–2: Intake: suction stroke 2–3: Isentropic Compression stroke 3–4: Heat addition stroke 4–5: Exhaust stroke (Isentropic expansion) 5–2: Heat rejection The distance between points 1–2 is the stroke of the engine. By dividing V2/V1, we get: r, where r is called the compression ratio of the engine.

Five-Stroke Engine

In 1879, Nikolaus Otto manufactured and sold a double expansion engine (the double and triple expansion principles had ample usage in steam engines), with two small cylinders at both sides of a low-pressure larger cylinder, where a second expansion of exhaust stroke gas took place; the owner returned it, alleging poor performance. In 1906, the concept was incorporated in a car built by EHV (Eisenhuth Horseless Vehicle Company) CT, USA; and in the 21st century Ilmor designed and successfully tested a 5-stroke double expansion internal combustion engine, with high power output and low SFC (Specific Fuel Consumption).

Six-Stroke Engine

The six-stroke engine was invented in 1883. Four kinds of six-stroke use a regular piston in a regular cylinder (Griffin six-stroke, Bajulaz six-stroke, Velozeta six-stroke and Crower six-stroke), firing every three crankshaft revolutions. The systems capture the wasted heat of the four-stroke Otto cycle with an injection of air or water.

The Beare Head and "piston charger" engines operate as opposed-piston engines, two pistons in a single cylinder, firing every two revolutions rather more like a regular four-stroke.

Other Cycles

The very first internal combustion engines did not compress the mixture. The first part of the piston downstroke drew in a fuel-air mixture, then the inlet valve closed and, in the remainder of the down-stroke, the fuel-air mixture fired. The exhaust valve opened for the piston upstroke. These attempts at imitating the principle of a steam engine were very inefficient. There are a number of variations of these cycles, most notably the Atkinson and Miller cycles. The diesel cycle is somewhat different.

Split-cycle engines separate the four strokes of intake, compression, combustion and exhaust into two separate but paired cylinders. The first cylinder is used for intake and compression. The compressed air is then transferred through a crossover passage from the compression cylinder into the second cylinder, where combustion and exhaust occur. A split-cycle engine is really an air compressor on one side with a combustion chamber on the other.

Previous split-cycle engines have had two major problems—poor breathing (volumetric efficiency) and low thermal efficiency. However, new designs are being introduced that seek to address these problems.

The Scuderi Engine addresses the breathing problem by reducing the clearance between the piston and the cylinder head through various turbo charging techniques. The Scuderi design requires the use of outwardly opening valves that enable the piston to move very close to the cylinder head without the interference of the valves. Scuderi addresses the low thermal efficiency via firing after top dead centre (ATDC).

Firing ATDC can be accomplished by using high-pressure air in the transfer passage to create sonic flow and high turbulence in the power cylinder.

Combustion Turbines

Jet Engine

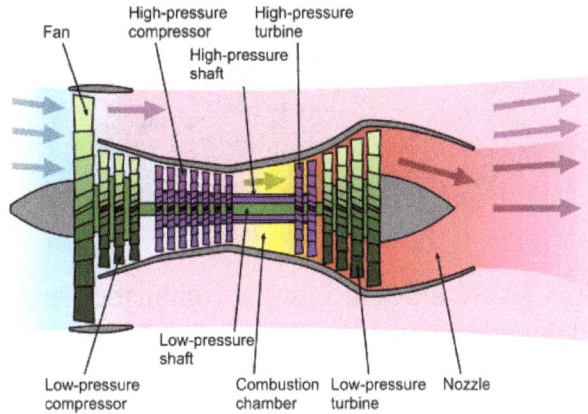

Turbofan Jet Engine

Jet engines use a number of rows of fan blades to compress air which then enters a combustor where it is mixed with fuel (typically JP fuel) and then ignited. The burning of the fuel raises the temperature of the air which is then exhausted out of the engine creating thrust. A modern turbofan engine can operate at as high as 48% efficiency.

There are six sections to a Fan Jet engine:

- Fan
- Compressor
- Combustor
- Turbine
- Mixer
- Nozzle

Gas Turbines

A gas turbine compresses air and uses it to turn a turbine. It is essentially a Jet engine which directs it's output to a shaft. There are three stages to a turbine: 1) air is drawn through a compressor where the temperature rises due to compression, 2) fuel is added in the combuster, and 3) hot air is exhausted through turbines blades which rotate a shaft connected to the compressor.

A gas turbine is a rotary machine similar in principle to a steam turbine and it consists of three main components: a compressor, a combustion chamber, and a turbine. The air, after being compressed in the compressor, is heated by burning fuel in it. About ⅔ of the heated air, combined with the products of combustion, expands in a turbine, producing work output that drives the compressor. The rest (about ⅓) is available as useful work output.

Turbine Power Plant

Gas Turbines are among the MOST efficient internal combustion engines. The General Electric 7HA and 9HA turbine electrical plants are rated at over 61% efficiency.

Brayton Cycle

A gas turbine is a rotary machine somewhat similar in principle to a steam turbine. It consists of three main components: compressor, combustion chamber, and turbine. The air is compressed by the compressor where a temperature rise occurs. The compressed air is further heated by combustion of injected fuel in the combustion chamber which expands the air. This energy rotates the turbine which powers the compressor via a mechanical coupling. The hot gases are then exhausted to provide thrust.

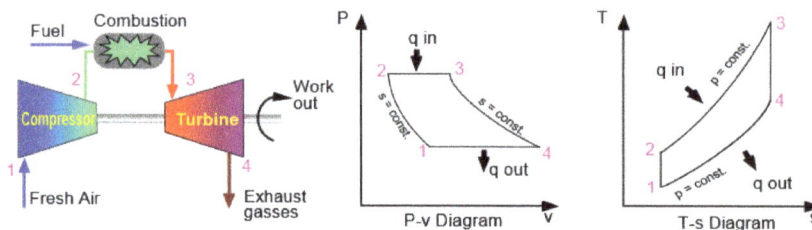
Brayton cycle

Gas turbine cycle engines employ a continuous combustion system where compression, combustion, and expansion occur simultaneously at different places in the engine—giving continuous power. Notably, the combustion takes place at constant pressure, rather than with the Otto cycle, constant volume.

Wankel Engines

The Wankel engine (rotary engine) does not have piston strokes. It operates with the same separation of phases as the four-stroke engine with the phases taking place in separate locations in the engine. In thermodynamic terms it follows the Otto engine cycle, so may be thought of as a "four-phase" engine. While it is true that three power strokes typically occur per rotor revolution, due to the 3:1 revolution ratio of the rotor to the eccentric shaft, only one power stroke per shaft revolution actually occurs. The drive (eccentric) shaft rotates once during every power stroke instead of twice (crankshaft), as in the Otto cycle, giving it a greater power-to-weight ratio than piston engines. This type of engine was most notably used in the Mazda RX-8, the earlier RX-7, and other vehicle models. The engine is also use in unmanned aerial vehicles, where the small size and weight and the high power-to-weight ratio are advantages.

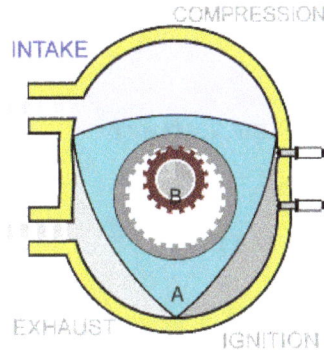

The Wankel rotary cycle. The shaft turns three times for each rotation of the rotor around the lobe and once for each orbital revolution around the eccentric shaft.

Forced Induction

Forced induction is the process of delivering compressed air to the intake of an internal combustion engine. A forced induction engine uses a gas compressor to increase the pressure, temperature and density of the air. An engine without forced induction is considered a naturally aspirated engine.

Forced induction is used in the automotive and aviation industry to increase engine power and efficiency. It particularly helps aviation engines, as they need to operate at high altitude.

Forced induction is achieved by a supercharger, where the compressor is directly powered from the engine shaft or, in the turbocharger, from a turbine powered by the engine exhaust.

Fuels and Oxidizers

All internal combustion engines depend on combustion of a chemical fuel, typically with oxygen from the air (though it is possible to inject nitrous oxide to do more of the same thing and gain a power boost). The combustion process typically results in the production of a great quantity of heat, as well as the production of steam and carbon dioxide and other chemicals at very high temperature; the temperature reached is determined by the chemical make up of the fuel and oxidisers, as well as by the compression and other factors.

Fuels

The most common modern fuels are made up of hydrocarbons and are derived mostly from fossil fuels (petroleum). Fossil fuels include diesel fuel, gasoline and petroleum gas, and the rarer use of propane. Except for the fuel delivery components, most internal combustion engines that are designed for gasoline use can run on natural gas or liquefied petroleum gases without major modifications. Large diesels can run with air mixed with gases and a pilot diesel fuel ignition injection.

Liquid and gaseous biofuels, such as ethanol and biodiesel (a form of diesel fuel that is produced from crops that yield triglycerides such as soybean oil), can also be used. Engines with appropriate modifications can also run on hydrogen gas, wood gas, or charcoal gas, as well as from so-called producer gas made from other convenient biomass. Experiments have also been conducted using powdered solid fuels, such as the magnesium injection cycle.

Presently, fuels used include:

- Petroleum:
    - Petroleum spirit (North American term: gasoline, British term: petrol)
    - Petroleum diesel.
    - Autogas (liquified petroleum gas).
    - Compressed natural gas.
    - Jet fuel (aviation fuel)
    - Residual fuel
- Coal:
    - Gasoline can be made from carbon (coal) using the Fischer-Tropsch process
    - Diesel fuel can be made from carbon using the Fischer-Tropsch process
- Biofuels and vegetable oils:
    - Peanut oil and other vegetable oils.
    - Woodgas, from an onboard wood gasifier using solid wood as a fuel
    - Biofuels:
        - Biobutanol (replaces gasoline).
        - Biodiesel (replaces petrodiesel).
        - Dimethyl Ether (replaces petrodiesel).
        - Bioethanol and Biomethanol (wood alcohol) and other biofuels (Flexible-fuel vehicle).
        - Biogas
- Hydrogen (mainly spacecraft rocket engines)

Even fluidized metal powders and explosives have seen some use. Engines that use gases for fuel are called gas engines and those that use liquid hydrocarbons are called oil engines; however, gasoline engines are also often colloquially referred to as, "gas engines" ("petrol engines" outside North America).

The main limitations on fuels are that it must be easily transportable through the fuel system to the combustion chamber, and that the fuel releases sufficient energy in the form of heat upon combustion to make practical use of the engine.

Diesel engines are generally heavier, noisier, and more powerful at lower speeds than gasoline engines. They are also more fuel-efficient in most circumstances and are used in heavy road vehicles, some automobiles (increasingly so for their increased fuel efficiency over gasoline engines), ships, railway locomotives, and light aircraft. Gasoline engines are used in most other road vehicles including most cars, motorcycles, and mopeds. Note that in Europe, sophisticated diesel-engined cars have taken over about 45% of the market since the 1990s. There are also engines that run on hydrogen, methanol, ethanol, liquefied petroleum gas (LPG), biodiesel, paraffin and tractor vaporizing oil (TVO).

Hydrogen

Hydrogen could eventually replace conventional fossil fuels in traditional internal combustion engines. Alternatively fuel cell technology may come to deliver its promise and the use of the internal combustion engines could even be phased out.

Although there are multiple ways of producing free hydrogen, those methods require converting combustible molecules into hydrogen or consuming electric energy. Unless that electricity is produced from a renewable source—and is not required for other purposes— hydrogen does not solve any energy crisis. In many situations the disadvantage of hydrogen, relative to carbon fuels, is its storage. Liquid hydrogen has extremely low density (14 times lower than water) and requires extensive insulation—whilst gaseous hydrogen requires heavy tankage. Even when liquefied, hydrogen has a higher specific energy but the volumetric energetic storage is still roughly five times lower than gasoline. However, the energy density of hydrogen is considerably higher than that of electric batteries, making it a serious contender as an energy carrier to replace fossil fuels. The 'Hydrogen on Demand' process creates hydrogen as needed, but has other issues, such as the high price of the sodium borohydride that is the raw material.

Oxidizers

One-cylinder gasoline engine, c. 1910

Since air is plentiful at the surface of the earth, the oxidizer is typically atmospheric oxygen, which has the advantage of not being stored within the vehicle. This increases the power-to-weight and power-to-volume ratios. Other materials are used for special purposes, often to increase power output or to allow operation under water or in space.

- Compressed air has been commonly used in torpedoes.

- Compressed oxygen, as well as some compressed air, was used in the Japanese Type 93 torpedo. Some submarines carry pure oxygen. Rockets very often use liquid oxygen.

- Nitromethane is added to some racing and model fuels to increase power and control combustion.

- Nitrous oxide has been used—with extra gasoline—in tactical aircraft, and in specially equipped cars to allow short bursts of added power from engines that otherwise run on gasoline and air. It is also used in the Burt Rutan rocket spacecraft.

- Hydrogen peroxide power was under development for German World War II submarines. It may have been used in some non-nuclear submarines, and was used on some rocket engines (notably the Black Arrow and the Me-163 rocket plane).

- Other chemicals such as chlorine or fluorine have been used experimentally, but have not been found practical.

Cooling

Cooling is required to remove excessive heat — over heating can cause engine failure, usually from wear(due to heat-induced failure of lubrication), cracking or warping. Two most common forms of engine cooling are air-cooled and water-cooled. Most modern automotive engines are both water and air-cooled, as the water/liquid-coolant is carried to air-cooled fins and/or fans, whereas larger engines may be singularly water-cooled as they are stationary and have a constant supply of water through water-mains or fresh-water, while most power tool engines and other small engines are air-cooled. Some engines (air or water-cooled) also have an oil cooler. In some engines, especially for turbine engine blade cooling and liquid rocket engine cooling, fuel is used as a coolant, as it is simultaneously preheated before injecting it into a combustion chamber.

Starting

Electric Starter as used in automobiles

Internal Combustion engines must have their cycles started. In reciprocating engines this is accomplished by turning the crankshaft (Wankel Rotor Shaft) which induces the cycles of intake, compression, combustion, and exhaust. The first engines were started with a turn of their fly-wheels, while the first vehicle (the Daimler Reitwagen) was started with a hand crank. All ICE engined automobiles were started with hand cranks until Charles Kettering developed the electric starter for automobiles.

The most often found methods of starting ICE today is with an electric motor. As diesel engines have become larger another method has come into use as well, that is Air Starters.

Another method of starting is to use compressed air that is pumped into some cylinders of an engine to start it turning.

With two wheeled vehicles their engines may be started in three ways:

- By pedaling, as on a bicycle
- By pushing the vehicle and then engaging the clutch (Run and Bump Starting)
- Electric Starting

There are also starters where a spring is compressed by a crank motion and then used to start an engine. Small engines use a pull rope mechanism called recoil starting as the rope returns to storage after it has been pulled fully out to start the engine.

Turbine engines are frequently started by electric motor, or by air.

Measures of Engine Performance

Engine types vary greatly in a number of different ways:

- energy efficiency
- fuel/propellant consumption (brake specific fuel consumption for shaft engines, thrust specific fuel consumption for jet engines)
- power-to-weight ratio
- thrust to weight ratio
- Torque curves (for shaft engines) thrust lapse (jet engines)
- Compression ratio for piston engines, overall pressure ratio for jet engines and gas turbines

Energy Efficiency

Once ignited and burnt, the combustion products—hot gases—have more available thermal energy than the original compressed fuel-air mixture (which had higher chemical energy). The available energy is manifested as high temperature and pressure that can be translated into work by the engine. In a reciprocating engine, the high-pressure gases inside the cylinders drive the engine's pistons.

Once the available energy has been removed, the remaining hot gases are vented (often by opening a valve or exposing the exhaust outlet) and this allows the piston to return to its previous position (top dead center, or TDC). The piston can then proceed to the next phase of its cycle, which varies between engines. Any heat that is not translated into work is normally considered a waste product and is removed from the engine either by an air or liquid cooling system.

Internal combustion engines are heat engines, and as such their theoretical efficiency can be approximated by idealized thermodynamic cycles. The thermal efficiency of a theoretical cycle cannot exceed that of the Carnot cycle, whose efficiency is determined by the difference between the lower and upper operating temperatures of the engine. The upper operating temperature of an engine is limited by two main factors; the thermal operating limits of the materials, and the auto-ignition resistance of the fuel. All metals and alloys have a thermal operating limit, and there is significant research into ceramic materials that can be made with greater thermal stability and desirable structural properties. Higher thermal stability allows for a greater temperature difference between the lower (ambient) and upper operating temperatures, hence greater thermodynamic efficiency. Also, as the cylinder temperature rises, the engine becomes more prone to auto-ignition. This is caused when the cylinder temperature nears the flash point of the charge. At this point, ignition can spontaneously occur before the spark plug fires, causing excessive cylinder pressures. Auto-ignition can be mitigated by using fuels with high auto-ignition resistance (octane rating), however it still puts an upper bound on the allowable peak cylinder temperature.

The thermodynamic limits assume that the engine is operating under ideal conditions: a frictionless world, ideal gases, perfect insulators, and operation for infinite time. Real world applications introduce complexities that reduce efficiency. For example, a real engine runs best at a specific load, termed its power band. The engine in a car cruising on a highway is usually operating significantly below its ideal load, because it is designed for the higher loads required for rapid acceleration. In addition, factors such as wind resistance reduce overall system efficiency. Engine fuel economy is measured in miles per gallon or in liters per 100 kilometres. The volume of hydrocarbon assumes a standard energy content.

Most iron engines have a thermodynamic limit of 37%. Even when aided with turbochargers and stock efficiency aids, most engines retain an *average* efficiency of about 18%-20 %. The latest technologies in Formula One engines have seen a boost in thermal efficiency to almost 47%. Rocket engine efficiencies are much better, up to 70%, because they operate at very high temperatures and pressures and can have very high expansion ratios. Electric motors are better still, at around 85 -90 % efficiency or more, but they rely on an external power source (often another heat engine at a power plant subject to similar thermodynamic efficiency limits). However large stationary power plant turbines are typically significantly more efficient and cleaner than small mobile combustion engines in vehicles.

There are many inventions aimed at increasing the efficiency of IC engines. In general, practical engines are always compromised by trade-offs between different properties such as efficiency, weight, power, heat, response, exhaust emissions, or noise. Sometimes economy also plays a role in not only the cost of manufacturing the engine itself, but also manufacturing and distributing the fuel. Increasing the engine's efficiency brings better fuel economy but only if the fuel cost per energy content is the same.

Measures of Fuel Efficiency and Propellant Efficiency

For stationary and shaft engines including propeller engines, fuel consumption is measured by calculating the brake specific fuel consumption, which measures the mass flow rate of fuel consumption divided by the power produced.

For internal combustion engines in the form of jet engines, the power output varies drastically with airspeed and a less variable measure is used: thrust specific fuel consumption (TSFC), which is the mass of propellant needed to generate impulses that is measured in either pound force-hour or the grams of propellant needed to generate an impulse that measures one kilonewton-second.

For rockets, TSFC can be used, but typically other equivalent measures are traditionally used, such as specific impulse and effective exhaust velocity.

Air and Noise Pollution

Air pollution

Internal combustion engines such as reciprocating internal combustion egines produce air polution emissions, due to incomplete combustion of c bonaceous fuel. The main derivatives of the process are carbon dioxide CO_2, water and some soot — also called particulate matter (PM). The effects of inhaling particulate matter have been studied in humans and animals and include asthma, lung cancer, cardiovascular issues, and premature death. There are, however, some additional products of the combustion process that include nitrogen oxides and sulfur and some uncombusted hydrocarbons, depending on the operating conditions and the fuel-air ratio.

Not all of the fuel is completely consumed by the combustion process; a small amount of fuel is present after combustion, and some of it reacts to form oxygenates, such as formaldehyde or acetaldehyde, or hydrocarbons not originally present in the input fuel mixture. Incomplete combustion usually results from insufficient oxygen to achieve the perfect stoichiometric ratio. The flame is "quenched" by the relatively cool cylinder walls, leaving behind unreacted fuel that is expelled with the exhaust. When running at lower speeds, quenching is commonly observed in diesel (compression ignition) engines that run on natural gas. Quenching reduces efficiency and increases knocking, sometimes causing the engine to stall. Incomplete combustion also leads to the production of carbon monoxide (CO). Further chemicals released are benzene and 1,3-butadiene that are also hazardous air pollutants.

Increasing the amount of air in the engine reduces emissions of incomplete combustion products, but also promotes reaction between oxygen and nitrogen in the air to produce nitrogen oxides (NOx). NOx is hazardous to both plant and animal health, and leads to the production of ozone (O_3). Ozone is not emitted directly; rather, it is a secondary air pollutant, produced in the atmosphere by the reaction of NOx and volatile organic compounds in the presence of sunlight. Ground-level ozone is harmful to human health and the environment. Though the same chemical substance, ground-level ozone should not be confused with stratospheric ozone, or the ozone layer, which protects the earth from harmful ultraviolet rays.

Carbon fuels contain sulfur and impurities that eventually produce sulfur monoxides (SO) and sulfur dioxide (SO_2) in the exhaust, which promotes acid rain.

In the United States, nitrogen oxides, PM, carbon monoxide, sulphur dioxide, and ozone, are regulated as criteria air pollutants under the Clean Air Act to levels where human health and welfare are protected. Other pollutants, such as benzene and 1,3-butadiene, are regulated as hazardous air pollutants whose emissions must be lowered as much as possible depending on technological and practical considerations.

NOx, carbon monoxide and other pollutants are frequently controlled via exhaust gas recirculation which returns some of the exhaust back into the engine intake, and catalytic converters, which convert exhaust chemicals to harmless chemicals.

Non-Road Engines

The emission standards used by many countries have special requirements for non-road engines which are used by equipment and vehicles that are not operated on the public roadways. The standards are separated from the road vehicles.

Noise Pollution

Significant contributions to noise pollution are made by internal combustion engines. Automobile and truck traffic operating on highways and street systems produce noise, as do aircraft flights due to jet noise, particularly supersonic-capable aircraft. Rocket engines create the most intense noise.

Idling

Internal combustion engines continue to consume fuel and emit pollutants when idling so it is desirable to keep periods of idling to a minimum. Many bus companies now instruct drivers to switch off the engine when the bus is waiting at a terminal.

In England, the Road Traffic Vehicle Emissions Fixed Penalty Regulations 2002 (Statutory Instrument 2002 No. 1808) introduced the concept of a *"stationary idling offence"*. This means that a driver can be ordered *"by an authorised person ... upon production of evidence of his authorisation, require him to stop the running of the engine of that vehicle"* and a *"person who fails to comply ... shall be guilty of an offence and be liable on summary conviction to a fine not exceeding level 3 on the standard scale"*. Only a few local authorities have implemented the regulations, one of them being Oxford City Council.

References

- Pulkrabek, Willard W. (1997). Engineering Fundamentals of the Internal Combustion Engine. Prentice Hall. p. 2. ISBN 9780135708545

- "Hand Cranking the Engine". Automobile in American Life and Society. University of Michigan-Dearborn. Retrieved 2016-09-01.

- Takaishi, Tatsuo; Numata, Akira; Nakano, Ryouji; Sakaguchi, Katsuhiko (March 2008). "Approach to High Efficiency Diesel and Gas Engines" (PDF). Mitsubishi Heavy Industries Technical Review. 45 (1). Retrieved 2011-02-04.

- "The Road Traffic (Vehicle Emissions) (Fixed Penalty) (England) Regulations 2002". 195.99.1.70. 2010-07-16. Retrieved 2010-08-28.

Methodologies used in Gas Turbines

This chapter discusses the methods of gas turbines in a critical manner by providing key analysis of the subject matter. Topics explored in this chapter include air-start system, axial compressor, turbine inlet air cooling and overall pressure ratio. This chapter discusses the methods of gas turbines in a critical manner providing key analysis to the subject matter.

Air-start System

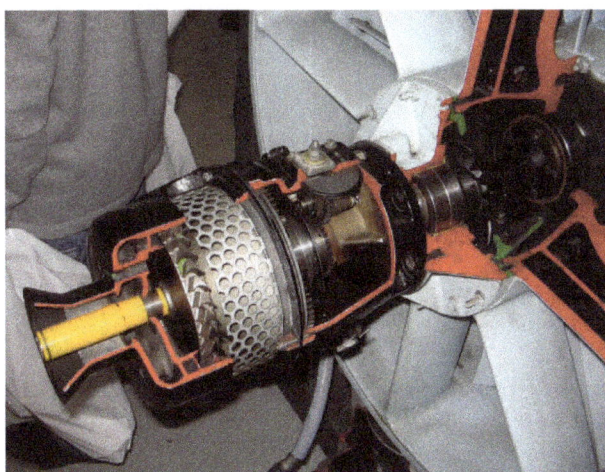

Cutaway of an air-start system of a General Electric J79 turbojet. The small turbine (next to yellow shaft) and epicyclic gearing (to right of perforated metal screen) are clearly visible.

An air-start system is a power source used to provide the initial rotation to start large diesel and gas turbine engines.

Diesel Engines

Direct Starting

Compared to a gasoline (petrol) engine, a Diesel engine has a very high compression ratio, an essential design feature, as it is the heat of compression that ignites the fuel. An electric starter with sufficient power to "crank" a large Diesel engine would itself be so large as to be impractical, thus the need for an alternative system.

When starting the engine, compressed air is admitted to whichever cylinder has a piston just over top dead center, forcing it downward. As the engine starts to turn, the air-start valve on the next cylinder in line opens to continue the rotation. After several rotations, fuel is injected into the cylinders, the engine starts running and the air is cut off.

To further complicate matters, a large engine is usually "blown over" first with zero fuel settings and the indicator cocks open, to prove that the engine is clear of any water build up and that everything is free to turn. After a successful blow ahead and a blow astern, the indicator cocks are closed on all the cylinders, and then the engine can be started on fuel. Significant complexity is added to the engine by using an air-start system, as the cylinder head must have an extra valve in each cylinder to admit the air in for starting, plus the required control systems. This added complexity and cost limits the use of air-starters to very large and expensive reciprocating engines.

Starter Motor

Another method of air-starting an internal combustion engine is by using compressed air or gas to drive a fluid motor in place of an electric motor. They can be used to start engines from 5 to 320 liters in size and if more starting power is necessary two or more motors can be used. Starters of this type are used in place of electric motors because of their lighter weight and higher reliability. They can also outlast an electric starter by a factor of three and are easier to rebuild.

Gas Turbines

An air-starter on a turbine engine would typically consist of a radial inward flow turbine, or axial flow turbine, which is connected to the High Pressure compressor spool through the accessory gearbox, plus the associated piping and valves. Compressed air is provided to the system by bleed air from the aircraft's auxiliary power unit or from an air compressor mounted on ground support equipment.

Advantages

Compared to electric starters, air-starters have a higher power-to-weight ratio. Electric starters and their wiring can become excessively hot if it takes longer than expected to start the engine, while air-starters can be run as long as their air supply lasts. Turbine starters are much simpler and are a natural fit for turbine engines, and thus are used extensively on large turbofan engines used on commercial and military aircraft.

Axial Compressor

An animated simulation of an axial compressor. The static blades are the stators.

An axial compressor is a machine that can continuously pressurise gases. It is a rotating, air-foil-based compressor in which the gas or working fluid principally flows parallel to the axis of rotation. This differs from other rotating compressors such as centrifugal compressors, axi-centrifugal compressors and mixed-flow compressors where the fluid flow will include a "radial component" through the compressor. The energy level of the fluid increases as it flows through the compressor due to the action of the rotor blades which exert a torque on the fluid. The stationary blades slow the fluid, converting the circumferential component of flow into pressure. Compressors are typically driven by an electric motor or a steam or a gas turbine.

Axial flow compressors produce a continuous flow of compressed gas, and have the benefits of high efficiency and large mass flow rate, particularly in relation to their size and cross-section. They do, however, require several rows of airfoils to achieve a large pressure rise, making them complex and expensive relative to other designs (e.g. centrifugal compressors).

Axial compressors are integral to the design of large gas turbines such as jet engines, high speed ship engines, and small scale power stations. They are also used in industrial applications such as large volume air separation plants, blast furnace air, fluid catalytic cracking air, and propane dehydrogenation. Due to high performance, high reliability and flexible operation during the flight envelope, they are also used in aerospace engines.

Typical application	Type of flow	Pressure ratio per stage	Efficiency per stage
Industrial	Subsonic	1.05–1.2	88–92%
Aerospace	Transonic	1.15–1.6	80–85%
Research	Supersonic	1.8–2.2	75–85%

Description

The compressor in a Pratt & Whitney TF30 turbofan engine.

Axial compressors consist of rotating and stationary components. A shaft drives a central drum, retained by bearings, which has a number of annular airfoil rows attached usually in pairs, one rotating and one stationary attached to a stationary tubular casing. A pair of rotating and stationary airfoils is called a stage. The rotating airfoils, also known as blades or rotors, accelerate the fluid. The stationary airfoils, also known as stators or vanes, convert the increased rotational kinetic energy into static pressure through diffusion and redirect the flow direction of the fluid, preparing it for the rotor blades of the next stage. The cross-sectional area between rotor drum and casing is reduced in the flow direction to maintain an optimum Mach number using variable geometry as the fluid is compressed.

Working

As the fluid enters and leaves in the axial direction, the centrifugal component in the energy equation does not come into play. Here the compression is fully based on diffusing action of the passages. The diffusing action in stator converts absolute kinetic head of the fluid into rise in pressure. The relative kinetic head in the energy equation is a term that exists only because of the rotation of the rotor. The rotor reduces the relative kinetic head of the fluid and adds it to the absolute kinetic head of the fluid i.e., the impact of the rotor on the fluid particles increases its velocity (absolute) and thereby reduces the relative velocity between the fluid and the rotor. In short, the rotor increases the absolute velocity of the fluid and the stator converts this into pressure rise. Designing the rotor passage with a diffusing capability can produce a pressure rise in addition to its normal functioning. This produces greater pressure rise per stage which constitutes a stator and a rotor together. This is the reaction principle in turbomachines. If 50% of the pressure rise in a stage is obtained at the rotor section, it is said to have a 50% reaction.

Design

The increase in pressure produced by a single stage is limited by the relative velocity between the rotor and the fluid, and the turning and diffusion capabilities of the airfoils. A typical stage in a commercial compressor will produce a pressure increase of between 15% and 60% (pressure ratios of 1.15–1.6) at design conditions with a polytropic efficiency in the region of 90–95%. To achieve different pressure ratios, axial compressors are designed with different numbers of stages and rotational speeds. As a rule of thumb we can assume that each stage in a given compressor has the same temperature rise (Delta T). Therefore, at the entry, temperature (Tstage) to each stage must increase progressively through the compressor and the ratio (Delta T)/(Tstage) entry must decrease, thus implying a progressive reduction in stage pressure ratio through the unit. Hence the rear stage develops a significantly lower pressure ratio than the first stage. Higher stage pressure ratios are also possible if the relative velocity between fluid and rotors is supersonic, but this is achieved at the expense of efficiency and operability. Such compressors, with stage pressure ratios of over 2, are only used where minimizing the compressor size, weight or complexity is critical, such as in military jets. The airfoil profiles are optimized and matched for specific velocities and turning. Although compressors can be run at other conditions with different flows, speeds, or pressure ratios, this can result in an efficiency penalty or even a partial or complete breakdown in flow (known as compressor stall and pressure surge respectively). Thus, a practical limit on the number of stages, and the overall pressure ratio, comes from the interaction of the different stages when required to work away from the design conditions. These "off-design" conditions can be mitigated to a certain extent by providing some flexibility in the compressor. This is achieved normally through the use of adjustable stators or with valves that can bleed fluid from the main flow between stages (inter-stage bleed). Modern jet engines use a series of compressors, running at different speeds;

to supply air at around 40:1 pressure ratio for combustion with sufficient flexibility for all flight conditions.

Kinetics and Energy Equations

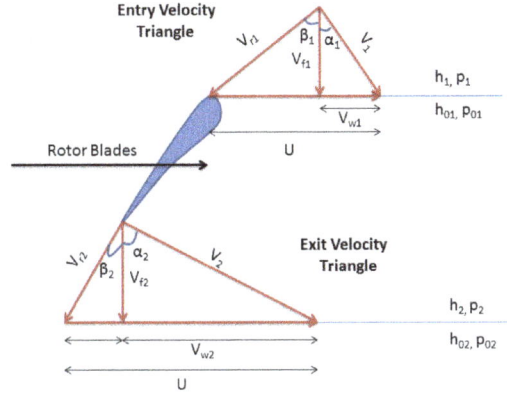

Velocity triangle of the swirling fluid entering and exiting the rotor blade

The law of moment of momentum states that the sum of the moments of external forces acting on a fluid which is temporarily occupying the control volume is equal to the net change of angular momentum flux through the control volume.

The swirling fluid enters the control volume at radius, r_1 , with tangential velocity, V_{w1} , and leaves at radius, r_2 , with tangential velocity, V_{w2} .

V_1 and V_2 are the absolute velocities at the inlet and outlet respectively.

V_{f1} and V_{f2} are the axial flow velocities at the inlet and outlet respectively.

V_{w1} and V_{w2} are the swirl velocities at the inlet and outlet respectively.

V_{r1} and V_{r2} are the blade-relative velocities at the inlet and outlet respectively.

U is the linear velocity of the blade.

α is the guide vane angle and β is the blade angle.

Rate of change of momentum, F is given by the equation:

$$F = \dot{m}\left(V_{w2} - V_{w1}\right) = \dot{m}\left(V_{f2}\tan\alpha_2 - V_{f1}\tan\alpha_1\right) \text{ (from velocity triangle)}$$

Power consumed by an ideal moving blade, P is given by the equation:

$$P = \dot{m}U\left(V_{f2}\tan\alpha_2 - V_{f1}\tan\alpha_1\right)$$

Change in enthalpy of fluid in moving blades:

$$P = \dot{m}\left(h_{02} - h_{01}\right) = \dot{m}c_p\left(T_{02} - T_{01}\right)$$

Therefore,

$$P = \dot{m}U\left(V_{f2}\tan\alpha_2 - V_{f1}\tan\alpha_1\right) = \dot{m}c_p\left(T_{02} - T_{01}\right)$$

which implies,

$$\delta(T_0)_{\text{isentropic}} = \frac{U}{c_p}\left(V_{f2}\tan\alpha_2 - V_{f1}\tan\alpha_1\right)$$

Isentropic compression in rotor blade,

$$p_2 - p_1 = p_1\left(\left[\frac{T_2}{T_1}\right]^{\frac{\gamma}{\gamma-1}} - 1\right)$$

Therefore,

$$\frac{(p_{02})_{\text{actual}}}{p_{01}} = \left(1 + \frac{\eta_{\text{stage}}\delta(T_0)_{\text{isentropic}}}{T_{01}}\right)^{\frac{\gamma}{\gamma-1}}$$

which implies

$$\frac{(p_{02})_{\text{actual}}}{p_{01}} = \left(1 + \frac{\eta_{\text{stage}}U}{T_{01}c_p}\left[V_{f2}\tan\alpha_2 - V_{f1}\tan\alpha_1\right]\right)^{\frac{\gamma}{\gamma-1}}$$

Degree of Reaction, The pressure difference between the entry and exit of the rotor blade is called reaction pressure. The change in pressure energy is calculated through *degree of reaction.*

$$R = \frac{h_2 - h_1}{h_{02} - h_{01}}$$

$$P = \dot{m}c_p\left(T_2 + \frac{V_2^2}{2c_p} - \left[T_1 + \frac{V_1^2}{2c_p}\right]\right)$$

$$P = \dot{m}\left(h_2 - h_1 + \left[\frac{V_2^2}{2} - \frac{V_1^2}{2}\right]\right)$$

$$h_2 - h_1 = \frac{V_{r1}^2}{2} - \frac{V_{r2}^2}{2}$$

$$T_2 - T_1 = \frac{V_{r1}^2}{2c_p} - \frac{V_{r2}^2}{2c_p}$$

Therefore,

$$R = \frac{V_{r1}^2 - V_{r2}^2}{V_{r1}^2 - V_{r2}^2 + V_1^2 - V_2^2}$$

Performance Characteristics

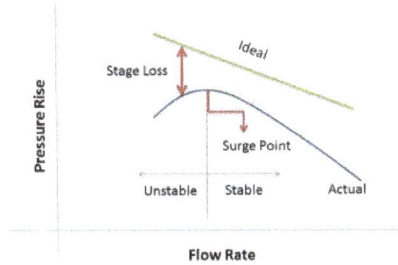

Reasons stating difference in ideal and actual performance curve in an axial compressor

A nonlinear model is developed to predict the transient response of a compression system subsequent to a perturbation from steady operating conditions. It is found that for the system investigated there is an important nondimensional parameter on which this response depends. Whether this parameter is above or below a critical value determines which mode of compressor instability, rotating stall or surge, will be encountered at the stall line. Representation of the performance characteristics of axial compressor can be done by following parameters:

- Pressure (P)

- Flow rate (Q)

- Non-dimensional flow rate ($\frac{\dot{m}\sqrt{T_{01}}}{P_{01}}$)

- Flow coefficient (ϕ)

- Stage loading coefficient ($\psi = \frac{V}{U^2}$)

By Plotting Graphs

Axial compressors, particularly near design conditions are, on the whole, amenable to analytical treatment, and usually a good estimate of their performance can be made before they are run. Away from the design points, the performances are conveniently thought of in terms of the overall characteristics of pressure-rises, temperature-rises, and efficiencies plotted against mass-flows.

- Pressure (P) as a function of flow rate (Q)

- Pressure ratio ($\frac{P_1}{P_2}$) as a function of non-dimensional flow rate

- Stage loading coefficient (ψ) as function of flow coefficient (ϕ)

we can determine performance of axial compressor

Difference between the ideal and actual curve arises due to stage loss. Stages losses in compressor are mainly due to blade friction, flow separation, unsteady flow and vane-blade spacing.

Pressure (P) as a function of flow rate (Q)

Off-Design Operation

The performance of a compressor is defined according to its design. But in actual practice, the operating point of the compressor deviates from the design- point which is known as off-design operation.

$$\psi = \phi(\tan \alpha_2 - \tan \alpha_1) \tag{1}$$

$$\tan \alpha_2 = \frac{1}{\phi} - \tan \beta_2 \tag{2}$$

from equation (1) and (2)

$$\psi = 1 - \phi(\tan \beta_2 + \tan \alpha_1)$$

The value of $(\tan \beta_2 + \tan \alpha_1)$ doesn't change for a wide range of operating points till stalling. Also $\alpha_1 = \alpha_3$ because of minor change in air angle at rotor and stator, where α_3 is diffuser blade angle.

$$J = \tan \beta_2 + \tan \alpha_3) \text{ is constant}$$

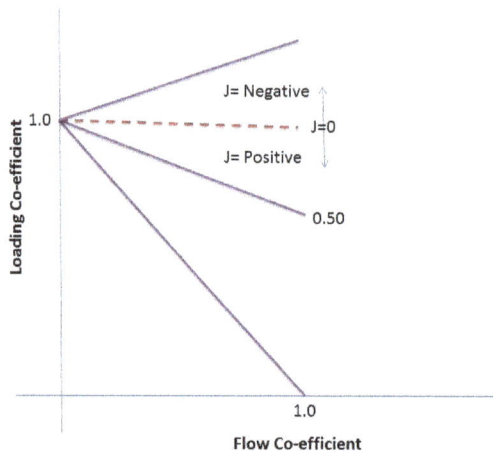

Off design characteristics curve of an axial compressor

Representing design values with (')

$$\psi' = 1 - J(\phi')$$

$$J = \frac{1 - \psi'}{\phi'} \tag{3}$$

for off-design operations (from eq. 3):

$$\psi = 1 - J(\phi)$$

$$\psi = 1 - \phi\left(\frac{1-\psi'}{\phi'}\right)$$

for positive values of J, slope of the curve is negative and vice versa.

Surging

In the plot of pressure-flow rate the line separating graph between two regions- unstable and stable is known as the surge line. This line is formed by joining surge points at different rpms. Unstable flow in axial compressors due to complete breakdown of the steady through flow is term as surging. This phenomenon affects the performance of compressor and is undesirable.

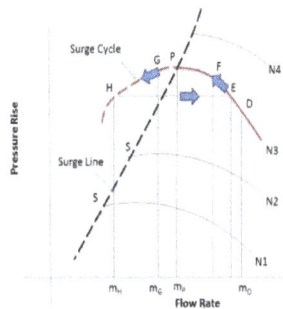

Various points on the performance curve depending upon the flow rates and pressure difference

Surge Cycle

Suppose the initial operating point D ($\dot{m}, P_D$) at some rpm N. On decreasing the flow- rate at same rpm along the characteristic curve by partial closing of the valve, the pressure in the pipe increases which will be taken care by increase in input pressure at the compressor. Further increase in pressure till point P (surge point), compressor pressure will increase. Further moving towards left keeping rpm constant, pressure in pipe will increase but compressor pressure will decrease leading to back air-flow towards the compressor. Due to this back flow, pressure in pipe will decrease because this unequal pressure condition cannot stay for a long period of time. Though valve position is set for lower flow rate say point G but compressor will work according to normal stable operation point say E, so path E-F-P-G-E will be followed leading to breakdown of flow, hence pressure in the compressor falls further to point H(P_H). This increase and decrease of pressure in pipe will occur repeatedly in pipe and compressor following the cycle E-F-P-G-H-E also known as the surge cycle.

This phenomenon will cause vibrations in the whole machine and may lead to mechanical failure. That is why left portion of the curve from the surge point is called unstable region and may cause damage to the machine. So the recommended operation range is on the right side of the surge line.

Stalling

Stalling is an important phenomenon that affects the performance of the compressor. An analysis is made of rotating stall in compressors of many stages, finding conditions under which a flow distortion can occur which is steady in a traveling reference frame, even though upstream total and downstream static pressure are constant. In the compressor, a pressure-rise hysteresis is assumed. It is a situation of separation of air flow at the aero-foil blades of the compressor. This phenomenon depending upon the blade-profile leads to reduced compression and drop in engine power.

Positive stalling

　　Flow separation occur on the suction side of the blade.

Negative stalling

　　Flow separation occur on the pressure side of the blade.

Negative stall is negligible compared to the positive stall because flow separation is least likely to occur on the pressure side of the blade.

In a multi-stage compressor, at the high pressure stages, axial velocity is very small. Stalling value decreases with a small deviation from the design point causing stall near the hub and tip regions whose size increases with decreasing flow rates. They grow larger at very low flow rate and affect the entire blade height. Delivery pressure significantly drops with large stalling which can lead to flow reversal. The stage efficiency drops with higher losses.

Rotating Stalling

Non-uniformity of air flow in the rotor blades may disturb local air flow in the compressor without upsetting it. The compressor continues to work normally but with reduced compression. Thus, rotating stall decreases the effectiveness of the compressor.

In a rotor with blades moving say towards right. Let some blades receives flow at higher incidence, this blade will stop positively. It creates obstruction in the passage between the blade to its left and itself. Thus the left blade will receive the flow at higher incidence and the blade to its right with decreased incidence. The left blade will experience more stall while the blade to its right will experience lesser stall. Towards the right stalling will decrease whereas it will increase towards its left. Movement of the rotating stall can be observed depending upon the chosen reference frame.

Effects

This reduces efficiency of the compressor

Forced vibrations in the blades due to passage through stall compartment.

These forced vibrations may match with the natural frequency of the blades causing resonance and hence failure of the blade.

Development

Early axial compressors offered poor efficiency, so poor that in the early 1920s a number of papers claimed that a practical jet engine would be impossible to construct. Things changed after A. A. Griffith published a seminal paper in 1926, noting that the reason for the poor performance was that existing compressors used flat blades and were essentially "flying stalled". He showed that the use of airfoils instead of the flat blades would increase efficiency to the point where a practical jet engine was a real possibility. He concluded the paper with a basic diagram of such an engine, which included a second turbine that was used to power a propeller.

Although Griffith was well known due to his earlier work on metal fatigue and stress measurement, little work appears to have started as a direct result of his paper. The only obvious effort was a test-bed compressor built by Hayne Constant, Griffith's colleague at the Royal Aircraft Establishment. Other early jet efforts, notably those of Frank Whittle and Hans von Ohain, were based on the more robust and better understood centrifugal compressor which was widely used in superchargers. Griffith had seen Whittle's work in 1929 and dismissed it, noting a mathematical error, and going on to claim that the frontal size of the engine would make it useless on a high-speed aircraft.

Real work on axial-flow engines started in the late 1930s, in several efforts that all started at about the same time. In England, Hayne Constant reached an agreement with the steam turbine company Metropolitan-Vickers (Metrovick) in 1937, starting their turboprop effort based on the Griffith design in 1938. In 1940, after the successful run of Whittle's centrifugal-flow design, their effort was re-designed as a pure jet, the Metrovick F.2. In Germany, von Ohain had produced several working centrifugal engines, some of which had flown including the world's first jet aircraft (He 178), but development efforts had moved on to Junkers (Jumo 004) and BMW (BMW 003), which used axial-flow designs in the world's first jet fighter (Messerschmitt Me 262) and jet bomber (Arado Ar 234). In the United States, both Lockheed and General Electric were awarded contracts in 1941 to develop axial-flow engines, the former a pure jet, the latter a turboprop. Northrop also started their own project to develop a turboprop, which the US Navy eventually contracted in 1943. Westinghouse also entered the race in 1942, their project proving to be the only successful one of the US efforts, later becoming the J30.

By the 1950s every major engine development had moved on to the axial-flow type. As Griffith had originally noted in 1929, the large frontal size of the centrifugal compressor caused it to have higher drag than the narrower axial-flow type. Additionally the axial-flow design could improve its compression ratio simply by adding additional stages and making the engine slightly longer. In the centrifugal-flow design the compressor itself had to be larger in diameter, which was much more difficult to "fit" properly on the aircraft. On the other hand, centrifugal-flow designs remained much less complex (the major reason they "won" in the race to flying examples) and therefore have a role in places where size and streamlining are not so important. For this reason they remain a major solution for helicopter engines, where the compressor lies flat and can be built to any needed size without upsetting the streamlining to any great degree.

Axial-Flow Jet Engines

In the jet engine application, the compressor faces a wide variety of operating conditions. On the ground at takeoff the inlet pressure is high, inlet speed zero, and the compressor spun at a variety of speeds as the power is applied. Once in flight the inlet pressure drops, but the inlet speed increases (due to the forward motion of the aircraft) to recover some of this pressure, and the compressor tends to run at a single speed for long periods of time.

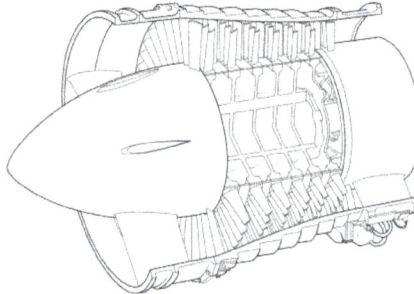

Low-pressure axial compressor scheme of the Olympus BOl.1 turbojet.

There is simply no "perfect" compressor for this wide range of operating conditions. Fixed geometry compressors, like those used on early jet engines, are limited to a design pressure ratio of about 4 or 5:1. As with any heat engine, fuel efficiency is strongly related to the compression ratio, so there is very strong financial need to improve the compressor stages beyond these sorts of ratios.

Additionally the compressor may stall if the inlet conditions change abruptly, a common problem on early engines. In some cases, if the stall occurs near the front of the engine, all of the stages from that point on will stop compressing the air. In this situation the energy required to run the compressor drops suddenly, and the remaining hot air in the rear of the engine allows the turbine to speed up the whole engine dramatically. This condition, known as surging, was a major problem on early engines and often led to the turbine or compressor breaking and shedding blades.

For all of these reasons, axial compressors on modern jet engines are considerably more complex than those on earlier designs.

Spools

All compressors have an optimum point relating rotational speed and pressure, with higher compressions requiring higher speeds. Early engines were designed for simplicity, and used a single large compressor spinning at a single speed. Later designs added a second turbine and divided the compressor into low-pressure and high-pressure sections, the latter spinning faster. This *two-spool* design, pioneered on the Bristol Olympus, resulted in increased efficiency. Further increases in efficiency may be realised by adding a third spool, but in practice the added complexity increases maintenance costs to the point of negating any economic benefit. That said, there are several three-spool engines in use, perhaps the most famous being the Rolls-Royce RB211, used on a wide variety of commercial aircraft.

Bleed Air, Variable Stators

As an aircraft changes speed or altitude, the pressure of the air at the inlet to the compressor will vary. In order to "tune" the compressor for these changing conditions, designs starting in the 1950s would "bleed" air out of the middle of the compressor in order to avoid trying to compress too much air in the final stages. This was also used to help start the engine, allowing it to be spun up without compressing much air by bleeding off as much as possible. Bleed systems were already commonly used anyway, to provide airflow into the turbine stage where it was used to cool the turbine blades, as well as provide pressurized air for the air conditioning systems inside the aircraft.

A more advanced design, the *variable stator*, used blades that can be individually rotated around their axis, as opposed to the power axis of the engine. For startup they are rotated to "closed", reducing compression, and then are rotated back into the airflow as the external conditions require. The General Electric J79 was the first major example of a variable stator design, and today it is a common feature of most military engines.

Closing the variable stators progressively, as compressor speed falls, reduces the slope of the surge (or stall) line on the operating characteristic (or map), improving the surge margin of the installed unit. By incorporating variable stators in the first five stages, General Electric Aircraft Engines has developed a ten-stage axial compressor capable of operating at a 23:1 design pressure ratio.

Design Notes

Energy Exchange Between Rotor and Fluid

The relative motion of the blades to the fluid adds velocity or pressure or both to the fluid as it passes through the rotor. The fluid velocity is increased through the rotor, and the stator converts kinetic energy to pressure energy. Some diffusion also occurs in the rotor in most practical designs.

The increase in velocity of the fluid is primarily in the tangential direction (swirl) and the stator removes this angular momentum.

The pressure rise results in a stagnation temperature rise. For a given geometry the temperature rise depends on the square of the tangential Mach number of the rotor row. Current turbofan engines have fans that operate at Mach 1.7 or more, and require significant containment and noise suppression structures to reduce blade loss damage and noise.

Compressor Maps

A map shows the performance of a compressor and allows determination of optimal operating conditions. It shows the mass flow along the horizontal axis, typically as a percentage of the design mass flow rate, or in actual units. The pressure rise is indicated on the vertical axis as a ratio between inlet and exit stagnation pressures.

A surge or stall line identifies the boundary to the left of which the compressor performance rapidly degrades and identifies the maximum pressure ratio that can be achieved for a given mass flow. Contours of efficiency are drawn as well as performance lines for operation at particular rotational speeds.

Compression Stability

Operating efficiency is highest close to the stall line. If the downstream pressure is increased beyond the maximum possible the compressor will stall and become unstable.

Typically the instability will be at the Helmholtz frequency of the system, taking the downstream plenum into account.

Turbine Inlet Air Cooling

An inlet air cooling system installed in a desert-dry area to increase turbine power output

Turbine inlet air cooling is a group of technologies and techniques consisting of cooling down the intake air of the gas turbine. The direct consequence of cooling the turbine inlet air is power output augmentation. It may also improve the energy efficiency of the system. This technology is widely used in hot climates with high ambient temperatures that usually coincides with on-peak demand period.

Principles

Gas turbines take in filtered, fresh ambient air and compress it in the compressor stage. The compressed air is mixed with fuel in the combustion chamber and ignited. This produces a high-temperature and high-pressure flow of exhaust gases that enter in a turbine and produce the shaft work output that is generally used to turn an electric generator as well as powering the compressor stage.

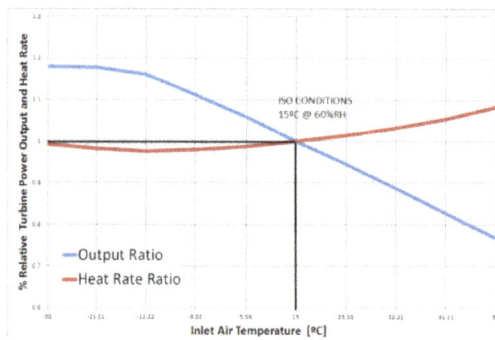

Effect of inlet air cooling on power output

As the gas turbine is a constant volume machine, the air volume introduced in the combustion chamber after the compression stage is fixed for a given shaft speed (rpm). Thus the air mass flow in is directly related to the density of air, and the introduced volume.

$$m = \rho V ,$$

where m is the mass, ρ is the density and V is the volume of the gas. As the volume V is fixed, only density ρ of the air can be modified to vary air mass. The density of the air depends of the relative humidity, altitude, pressure drop and temperature.

$$\rho_{humid\ air} = \frac{p_d}{R_d T} + \frac{p_v}{R_v T} = \frac{p_d M_d + p_v M_v}{RT}$$

where:

$\rho_{humid\ air}$ Density of the humid air (kg/m³)

p_d = Partial pressure of dry air (Pa)

R_d = Specific gas constant for dry air, 287.058 J/(kg·K)

T = Temperature (K)

p_v = Pressure of water vapor (Pa)

R_v = Specific gas constant for water vapor, 461.495 J/(kg·K)

M_d = Molar mass of dry air, 0.028964 (kg/mol)

M_v = Molar mass of water vapor, 0.018016 (kg/mol)

R = Universal gas constant, 8.314 J/(K·mol)

The performance of a gas turbine, its efficiency (heat rate) and the generated power output strongly depend on the climate conditions, which may decrease the output power ratings by up to 40%.

To operate the turbine at ISO conditions and recover performance, several inlet air cooling systems have been promoted.

Applied Technologies

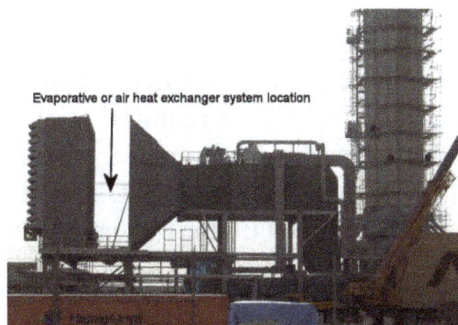

Filter-house modified to place the heat exchanger after the filtering stage.

Different technologies are available in the market. Each particular technology has its advantages and inconveniences according to different factors such as ambient conditions, investment cost and payback time, power output increase and cooling capacity.

Fogging

Inlet air fogging consists of spraying finely atomized water (fog) into the inlet airflow of a gas turbine engine. The water droplets evaporate quickly, which cools the air and increases the power output of the turbine.

Demineralized water is typically pressurized to 2000 psi (138 bar) then injected into the inlet air duct through an array of stainless steel fog nozzles. Demineralized water is used in order to prevent fouling of the compressor blades that would occur if water with mineral content were evaporated in the airflow. Fog systems typically produce a water spray, with about 90% of the water flow being in droplets that are 20 microns in diameter or smaller.

Inlet fogging has been in commercial use since the late 1980s and is a popular retrofit technology. As of 2015, there were more than 1000 inlet fog systems installed around the world. Inlet fog systems are, "simple, easy to install and operate" and less expensive than other power augmentation systems such as evaporative coolers and chillers.

Inlet fogging is the least expensive gas turbine inlet air cooling option and has low operating costs, particularly when one accounts for the fact that fog systems impose only a negligible pressure drop on the inlet airflow when compared to media-type evaporative coolers.

Fog nozzle manifolds are typically located in the inlet air duct just downstream of the final air filters but other locations can be desirable depending on the design of the inlet duct and the intended use of the fog system.

On a hot afternoon in a desert climate, it is possible to cool by as much as 40 °F (22.2 °C), while in a humid climate hot-afternoon cooling potential can be just 10 °F (5.6 °C) or less. Nevertheless, there are many successful inlet-fogging installations in humid climates such as Thailand, Malaysia and the American Gulf States.

Inlet fogging reduces emissions of Oxides of Nitrogen (NOx) because the additional water vapor quenches hot spots in the combustors of the gas turbine.

Wet Compression

Fog systems can be used to produce more power that can be obtained by evaporative cooling alone. This is accomplished by spraying more fog than is required to fully saturate the inlet air. The excess fog droplets are carried into the gas turbine compressor where they evaporate and produce an intercooling effect, which results in a further power boost. This technique was first employed on an experimental gas turbine in Norway in 1903. There are many successful systems in operation today.

Several gas turbine manufactures offer both fogging and wet compression systems. Systems are also available from third-party manufacturers.

Evaporative Cooling

The evaporative cooler is a wetted rigid media where water is distributed throughout the header and where air passes through the wet porous surface. Part of the water is evaporated, absorbing the sensible heat from the air and increasing its relative humidity. The air dry-bulb temperature is decreased but the wet-bulb temperature is not affected. Similar to the fogging system, the theoretical limit is the wet bulb temperature, but performance of the evaporative cooler is usually around 80%. Water consumption is less than that of fogging cooling.

Cooling systems based on latent heat as the water evaporates are preferred in dry/desert climates not near the sea where the relative humidity is low, and where the system can boost the turbine output by nearly 12%. The problem is that for a desert climate, a large amount of water is a limiting factor. For warm and humid climates the evaporative-kind of air cooling system may not increase the turbine output by more than 2-3%.

Vapour Compression Chiller

Turbine inlet air cooling filter-house modification to place the cooling coil coming from ammonia compression chiller plant

In a mechanical compression chiller technology, the coolant is circulated through a chilling coil heat exchanger that is inserted in the filter house, downstream from the filtering stage. Downstream from the coil, a droplet catcher is installed to collect moisture and water drops. The mechanical chiller can increase the turbine output and performance better than wetted technologies due to the fact that inlet air can be chilled below the wet bulb temperature, indifferent to the weather conditions. Compression chiller equipment has higher electricity consumption than evaporative systems. Initial capital cost is also higher, however turbine power augmentation and efficiency is maximized, and the extra-cost is amortized due to increased output power.

The majority of such systems involve more than one chiller unit and the configuration of the chillers can have a great bearing on the system parasitic power consumption. The series counter-flow configuration can reduce the compressor work needed on each chiller, improving the overall chiller system by as much as 8%.

Other options such a steam driven compression are also used in industry.

Vapour-Absorption Chiller

In vapor-absorption chillers technology, thermal energy is used to produce cooling instead of mechanical energy. The heat source is usually leftover steam coming from combined cycle, and it is bypassed to drive the cooling system. Compared to mechanical chillers, absorption chillers have a low coefficient of performance, however, it should be taken into consideration that this chiller usually uses waste heat, which decreases the operational cost.

Combination with Thermal Energy Storage

A thermal energy storage tank is a naturally stratified thermal accumulator that allows the storage of chilled water produced during off-peak time, to use this energy later during on-peak time to chill the turbine inlet air and increment its power output. A thermal energy storage tank reduces operational cost and refrigerant plant capacity. One advantage is the production of chilled water when demand is low, using the excess of power generation, which usually coincides with the night, when ambient temperature is low and chillers have better performance. Another advantage is the reduction of the chilling plant capacity and operational cost in comparison with an on-line chilling system, which produce delays during periods of low demand.

Benefits

In areas where there is demand cooling, daily summer on-peak periods coincide with the highest atmospheric temperatures, which may reduce the efficiency and power gas turbines. With the vapor mechanical compression technologies, cooling can be used during these periods so that the performance and the power output of the turbine may be less affected by ambient conditions

Another benefit is the lower cost per extra inlet-cooling kilowatt compared to newly installed gas turbine kilowatt. Moreover, the extra inlet-cooling kilowatt uses less fuel than the new turbine kilowatt due to the lower heat-rate (higher efficiency) of the chilled turbine. Other benefits may include the incrementation of steam mass flow in a combined cycle, the reduction of turbine emissions (SOx, NOx, CO2), and increase in power-to-installed volume ratio.

Calculating the benefits of turbine air cooling requires a study to determine payback periods, taking into consideration several aspects like ambient conditions, cost of water, hourly electric demand values, cost of fuel.

Overall pressure ratio

In aeronautical engineering, overall pressure ratio, or overall compression ratio, is the ratio of the stagnation pressure as measured at the front and rear of the compressor of a gas turbine engine. The terms *compression ratio* and *pressure ratio* are used interchangeably. Overall compression ratio also means the *overall cycle pressure ratio* which includes intake ram.

History of Overall Pressure Ratios

Early jet engines had limited pressure ratios due to construction inaccuracies of the compressors and various material limits. For instance, the Junkers Jumo 004 from World War II had an overall pressure ratio 3.14:1. The immediate post-war SNECMA Atar improved this marginally to 5.2:1. Improvements in materials, compressor blades, and especially the introduction of multi-spool engines with several different rotational speeds, led to the much higher pressure ratios common today. Modern civilian engines generally operate between 30 and 40:1. The three-spool Rolls-Royce Trent 900 used on the Airbus A380, for instance, has a pressure ratio of about 39:1.

Advantages of High Overall Pressure Ratios

Generally speaking, a higher overall pressure ratio implies higher efficiency, but the engine will usually weigh more, so there is a compromise. A high overall pressure ratio permits a larger area ratio nozzle to be fitted on the jet engine. This means that more of the heat energy is converted to jet speed, and energetic efficiency improves. This is reflected in improvements in the engine's specific fuel consumption.

Disadvantages of High Overall Pressure Ratios

One of the primary limiting factors on pressure ratio in modern designs is that the air heats up as it is compressed. As the air travels through the compressor stages it can reach temperatures that pose a material failure risk for the compressor blades. This is especially true for the last compressor stage, and the outlet temperature from this stage is a common figure of merit for engine designs.

Military engines are often forced to work under conditions that maximize the heating load. For instance, the General Dynamics F-111 was required to operate at speeds of Mach 1.1 at sea level. As a side-effect of these wide operating conditions, and generally older technology in most cases, military engines typically have lower overall pressure ratios. The Pratt & Whitney TF30 used on the F-111 had a pressure ratio of about 20:1, while newer engines like the General Electric F110 and Pratt & Whitney F135 have improved this to about 30:1.

An additional concern is weight. A higher compression ratio implies a heavier engine, which in turn costs fuel to carry around. Thus, for a particular construction technology and set of flight plans an optimal overall pressure ratio can be determined.

Examples

Engine	Overall pressure ratio	Major applications	Notes
Rolls-Royce Trent XWB	52:1	A350 XWB	
General Electric GE90	42:1	777	
General Electric CF6	30.5:1	747, 767, A300, MD-11, C-5	
General Electric F110	30:1	F-14, F-15, F-16	

Pratt & Whitney TF30	20:1	F-14, F-111	
Rolls-Royce/Snecma Olympus 593	15.5:1	Concorde	The Concorde's Olympus engines received additional compression from their supersonic air intakes, yielding an effective overall pressure ratio of 80:1.

Differences from other similar terms

The term should not be confused with the more familiar term compression ratio applied to reciprocating engines. Compression ratio is a ratio of volumes. In the case of the Otto cycle reciprocating engine, the maximum expansion of the charge is limited by the mechanical movement of the pistons (or rotor), and so the compression can be measured by simply comparing the volume of the cylinder with the piston at the top and bottom of its motion. The same is not true of the "open ended" gas turbine, where operational and structural conciderations are the limiting factors. Nevertheless the two terms are similar in that they both offer a quick way of determining overall efficiency relative to other engines of the same class.

The broadly equivalent measure of rocket engine efficiency is chamber pressure/exit pressure, and this ratio can be over 2000 for the Space Shuttle Main Engine.

Compression Ratio Versus Overall Pressure Ratio

For any given gas mix compression ratio and overall pressure ratio are interrelated as follows:

CR	1:1	3:1	5:1	10:1	15:1	20:1	25:1	35:1
PR	1:1	4:1	10:1	22:1	40:1	56:1	75:1	110:1

The reason for this difference is that compression ratio is defined via the volume reduction,

$$CR = \frac{V_1}{V_2},$$

Pressure ratio is defined as the pressure increase

$$PR = \frac{P_2}{P_1}.$$

From the combined gas law we get:

$$\frac{P_1 V_1}{T_1} = \frac{P_2 V_2}{T_2} \Rightarrow \frac{V_1}{V_2} = \frac{T_1}{T_2} \frac{P_2}{P_1} \Leftrightarrow CR = \frac{T_1}{T_2} PR$$

Since T_2 is much higher than T_1 (compressing gases puts work into them, i.e. heats them up), CR is much lower than PR.

Applications of Gas Turbines

Due to the extensive applications of gas turbines in the field of power generation, they are also employed in related fields. This chapter provides the reader with a thorough understanding of the applications of gas turbines in the fields of marine propulsion, trains, electric locomotives, prop-fans, aircrafts etc.

Marine Propulsion

A view of a ship's engine room

Marine propulsion is the mechanism or system used to generate thrust to move a ship or boat across water. While paddles and sails are still used on some smaller boats, most modern ships are propelled by mechanical systems consisting of an electric motor or engine turning a propeller, or less frequently, in pump-jets, an impeller. Marine engineering is the discipline concerned with the engineering design process of marine propulsion systems.

Marine steam engines were the first mechanical engines used in marine propulsion, however they have mostly been replaced by two-stroke or four-stroke diesel engines, outboard motors, and gas turbine engines on faster ships. Nuclear reactors producing steam are used to propel warships and icebreakers. Nuclear reactors to power commercial vessels has not been adopted by the marine industry. Electric motors using electric battery storage have been used for propulsion on submarines and electric boats and have been proposed for energy-efficient propulsion. Development in liquefied natural gas (LNG) fueled engines are gaining recognition for their low emissions and cost advantages. Stirling engines, which are more efficient, quieter, smoother running producing less harmful emissions than diesel engines, propel a number of small submarines. The Stirling engine has yet to be upscaled for larger surface ships.

Power sources

Pre-mechanisation

A wind propelled fishing boat in Mozambique

Until the application of the coal-fired steam engine to ships in the early 19th century, oars or the wind were used to assist watercraft propulsion. Merchant ships predominantly used sail, but during periods when naval warfare depended on ships closing to ram or to fight hand-to-hand, galley were preferred for their manoeuvrability and speed. The Greek navies that fought in the Peloponnesian War used triremes, as did the Romans at the Battle of Actium. The development of naval gunnery from the 16th century onward meant that manoeuvrability took second place to broadside weight; this led to the dominance of the sail-powered warship over the following three centuries.

In modern times, human propulsion is found mainly on small boats or as auxiliary propulsion on sailboats. Human propulsion includes the push pole, rowing, and pedals.

Propulsion by sail generally consists of a sail hoisted on an erect mast, supported by stays, and controlled by lines made of rope. Sails were the dominant form of commercial propulsion until the late nineteenth century, and continued to be used well into the twentieth century on routes where wind was assured and coal was not available, such as in the South American nitrate trade. Sails are now generally used for recreation and racing, although innovative applications of kites/royals, turbosails, rotorsails, wingsails, windmills and SkySails's own kite buoy-system have been used on larger modern vessels for fuel savings.

Reciprocating Steam Engines

The development of piston-engined steamships was a complex process. Early steamships were fueled by wood, later ones by coal or fuel oil. Early ships used stern or side paddle wheels, while later ones used screw propellers.

SS *Ukkopekka* uses a triple expansion steam engine

The first commercial success accrued to Robert Fulton's *North River Steamboat* (often called *Clermont*) in US in 1807, followed in Europe by the 45-foot *Comet* of 1812. Steam propulsion progressed considerably over the rest of the 19th century. Notable developments include the steam surface condenser, which eliminated the use of sea water in the ship's boilers. This, along with improvements in boiler technology, permitted higher steam pressures, and thus the use of higher efficiency multiple expansion (compound) engines. As the means of transmitting the engine's power, paddle wheels gave way to more efficient screw propellers.

Steam Turbines

Steam turbines were fueled by coal or, later, fuel oil or nuclear power. The marine steam turbine developed by Sir Charles Algernon Parsons raised the power-to-weight ratio. He achieved publicity by demonstrating it unofficially in the 100-foot *Turbinia* at the Spithead Naval Review in 1897. This facilitated a generation of high-speed liners in the first half of the 20th century, and rendered the reciprocating steam engine obsolete; first in warships, and later in merchant vessels.

In the early 20th century, heavy fuel oil came into more general use and began to replace coal as the fuel of choice in steamships. Its great advantages were convenience, reduced manpower by removal of the need for trimmers and stokers, and reduced space needed for fuel bunkers.

In the second half of the 20th century, rising fuel costs almost led to the demise of the steam turbine. Most new ships since around 1960 have been built with diesel engines. The last major passenger ship built with steam turbines was the *Fairsky*, launched in 1984. Similarly, many steam ships were re-engined to improve fuel efficiency. One high profile example was the 1968 built *Queen Elizabeth 2* which had her steam turbines replaced with a diesel-electric propulsion plant in 1986.

Most new-build ships with steam turbines are specialist vessels such as nuclear-powered vessels, and certain merchant vessels (notably Liquefied Natural Gas (LNG) and coal carriers) where the cargo can be used as bunker fuel.

LNG Carriers

New LNG carriers (a high growth area of shipping) continue to be built with steam turbines. The natural gas is stored in a liquid state in cryogenic vessels aboard these ships, and a small amount of 'boil off' gas is needed to maintain the pressure and temperature inside the vessels within operating limits. The 'boil off' gas provides the fuel for the ship's boilers, which provide steam for the turbines, the simplest way to deal with the gas. Technology to operate internal combustion engines (modified marine two-stroke diesel engines) on this gas has improved, however, such engines are starting to appear in LNG carriers; with their greater thermal efficiency, less gas is burnt. Developments have also been made in the process of re-liquifying 'boil off' gas, letting it be returned to the cryogenic tanks. The financial returns on LNG are potentially greater than the cost of the marine-grade fuel oil burnt in conventional diesel engines, so the re-liquefaction process is starting to be used on diesel engine propelled LNG carriers. Another factor driving the change from turbines to diesel engines for LNG carriers is the shortage of steam turbine qualified seagoing engineers. With the lack of turbine powered ships in other shipping sectors, and the rapid rise in size of the worldwide LNG fleet, not enough have been trained to meet the demand. It may be that the days are numbered for marine steam turbine propulsion systems, even though all but sixteen of the orders for new LNG carriers at the end of 2004 were for steam turbine propelled ships.

The NS *Savannah* was the first nuclear-powered cargo-passenger ship

Nuclear-Powered Steam Turbines

In these vessels, the nuclear reactor heats water to create steam to drive the turbines. Due to low prices of diesel oil, nuclear propulsion is rare except in some Navy and specialist vessels such as icebreakers. In large aircraft carriers, the space formerly used for ship's bunkerage could be used instead to bunker aviation fuel. In submarines, the ability to run submerged at high speed and in relative quiet for long periods holds obvious advantages. A few cruisers have also employed nuclear power; as of 2006, the only ones remaining in service are the Russian *Kirov* class. An example of a non-military ship with nuclear marine propulsion is the *Arktika* class icebreaker with 75,000 shaft horsepower (55,930 kW). Commercial experiments such as the NS *Savannah* have so far proved uneconomical compared with conventional propulsion.

In recent times, there is some renewed interest in commercial nuclear shipping. Nuclear-powered cargo ships could lower costs associated with carbon dioxide emissions and travel at higher cruise speeds than conventional diesel powered vessels.

Reciprocating diesel engines

Most modern ships use a reciprocating diesel engine as their prime mover, due to their operating simplicity, robustness and fuel economy compared to most other prime mover mechanisms.

The rotating crankshaft can be directly coupled to the propeller with slow speed engines, via a reduction gearbox for medium and high speed engines, or via an alternator and electric motor in diesel-electric vessels. The rotation of the crankshaft is connected to the camshaft or a hydraulic pump on an intelligent diesel.

A modern diesel engine aboard a cargo ship

The reciprocating marine diesel engine first came into use in 1903 when the diesel electric river-tanker *Vandal* was put into service by Branobel. Diesel engines soon offered greater efficiency than the steam turbine, but for many years had an inferior power-to-space ratio. The advent of turbo-charging however hastened their adoption, by permitting greater power densities.

Diesel engines today are broadly classified according to

- Their operating cycle: two-stroke engine or four-stroke engine

- Their construction: crosshead, trunk, or opposed piston

- Their speed

    - Slow speed: any engine with a maximum operating speed up to 300 revolutions per minute (rpm), although most large two-stroke slow speed diesel engines operate below 120 rpm. Some very long stroke engines have a maximum speed of around 80 rpm. The largest, most powerful engines in the world are slow speed, two stroke, crosshead diesels.

    - Medium speed: any engine with a maximum operating speed in the range 300-900 rpm. Many modern four-stroke medium speed diesel engines have a maximum operating speed of around 500 rpm.

    - High speed: any engine with a maximum operating speed above 900 rpm.

Most modern larger merchant ships use either slow speed, two stroke, crosshead engines, or medium speed, four stroke, trunk engines. Some smaller vessels may use high speed diesel engines.

The size of the different types of engines is an important factor in selecting what will be installed in a new ship. Slow speed two-stroke engines are much taller, but the footprint required is smaller than that needed for equivalently rated four-stroke medium speed diesel engines. As space above the waterline is at a premium in passenger ships and ferries (especially ones with a car deck), these ships tend to use multiple medium speed engines resulting in a longer, lower engine room than that needed for two-stroke diesel engines. Multiple engine installations also give redundancy in the event of mechanical failure of one or more engines, and the potential for greater efficiency over a wider range of operating conditions.

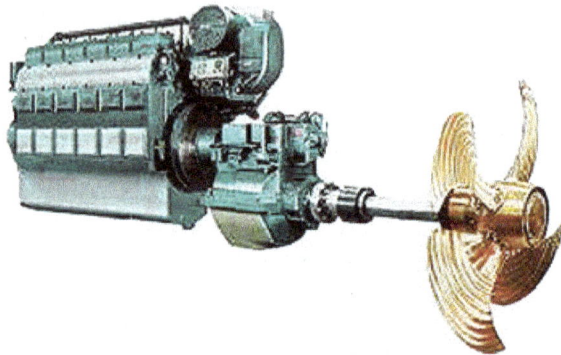

4-Stroke Marine Diesel Engine System

As modern ships' propellers are at their most efficient at the operating speed of most slow speed diesel engines, ships with these engines do not generally need gearboxes. Usually such propulsion systems consist of either one or two propeller shafts each with its own direct drive engine. Ships propelled by medium or high speed diesel engines may have one or two (sometimes more) propellers, commonly with one or more engines driving each propeller shaft through a gearbox. Where more than one engine is geared to a single shaft, each engine will most likely drive through a clutch, allowing engines not being used to be disconnected from the gearbox while others keep running. This arrangement lets maintenance be carried out while under way, even far from port.

LNG Engines

Shipping companies are required to comply with the International Maritime Organization (IMO) and the International Convention for the Prevention of Pollution from Ships emissions rules. Dual fuel engines are fueled by either marine grade diesel, heavy fuel oil, or liquefied natural gas (LNG). A Marine LNG Engine has multiple fuel options, allowing vessels to transit without relying on one type of fuel. Studies show that LNG is the most efficient of fuels, although limited access to LNG fueling stations limits the production of such engines. Vessels providing services in the LNG industry have been retrofitted with dual-fuel engines, and have been proved to be extremely effective. Benefits of dual-fuel engines include fuel and operational flexibility, high efficiency, low emissions, and operational cost advantages. Liquefied natural gas engines offer the marine transportation industry with an environmentally friendly alternative to provide power to vessels. In 2010, STX Finland and Viking Line signed an agreement to begin construction on what would be the largest environmentally friendly cruise ferry. Construction of NB 1376 will be completed in 2013. According to Viking Line, vessel NB 1376 will primarily be fueled by liquefied natural gas. Vessel NB 1376 nitrogen oxide emissions will be almost zero, and sulphur oxide emissions will be at least 80% below the International Maritime Organization's (IMO) standards. Company profits from tax cuts and operational cost advantages has led to the gradual growth of LNG fuel use in engines.

Gas Turbines

Combined marine propulsion
CODOG
CODAG
CODLAG
CODAD
COSAG
COGOG
COGAG
COGAS
CONAS
IEP or IFEP

Many warships built since the 1960s have used gas turbines for propulsion, as have a few passenger ships, like the jetfoil. Gas turbines are commonly used in combination with other types of engine. Most recently, the *RMS Queen Mary 2* has had gas turbines installed in addition to diesel engines. Because of their poor thermal efficiency at low power (cruising) output, it is common for ships using them to have diesel engines for cruising, with gas turbines reserved for when higher speeds are needed. However, in the case of passenger ships the main reason for installing gas turbines has been to allow a reduction of emissions in sensitive environmental areas or while in port. Some warships, and a few modern cruise ships have also used steam turbines to improve the efficiency of their gas turbines in a combined cycle, where waste heat from a gas turbine exhaust is utilized to boil water and create steam for driving a steam turbine. In such combined cycles, thermal efficiency can be the same or slightly greater than that of diesel engines alone; however, the grade of fuel needed for these gas turbines is far more costly than that needed for the diesel engines, so the running costs are still higher.

Stirling Engines

Since the late 1980s, Swedish shipbuilder Kockums has built a number of successful Stirling engine powered submarines. The submarines store compressed oxygen to allow more efficient and cleaner external fuel combustion when submerged, providing heat for the Stirling engine's operation. The engines are currently used on submarines of the *Gotland* and *Södermanland* classes. and the Japanese *Sōryū*-class submarine. These are the first submarines to feature Stirling air-independent propulsion (AIP), which extends the underwater endurance from a few days to several weeks.

The heat sink of a Stirling engine is typically the ambient air temperature. In the case of medium to high power Stirling engines, a radiator is generally required to transfer the heat from the engine to the ambient air. Stirling marine engines have the advantage of using the ambient temperature water. Placing the cooling radiator section in seawater rather than ambient air allows for the radiator to be smaller. The engine's cooling water may be used directly or indirectly for heating and cooling purposes of the ship. The Stirling engine has potential for surface-ship propulsion, as the engine's larger physical size is less of a concern.

Screws

Marine propellers are also known as "screws". There are many variations of marine screw systems, including twin, contra-rotating, controllable-pitch, and nozzle-style screws. While smaller vessels tend to have a single screw, even very large ships such as tankers, container ships and bulk carriers may have single screws for reasons of fuel efficiency. Other vessels may have twin, triple or quadruple screws. Power is transmitted from the engine to the screw by way of a propeller shaft, which may or may not be connected to a gearbox.

Paddle Wheels

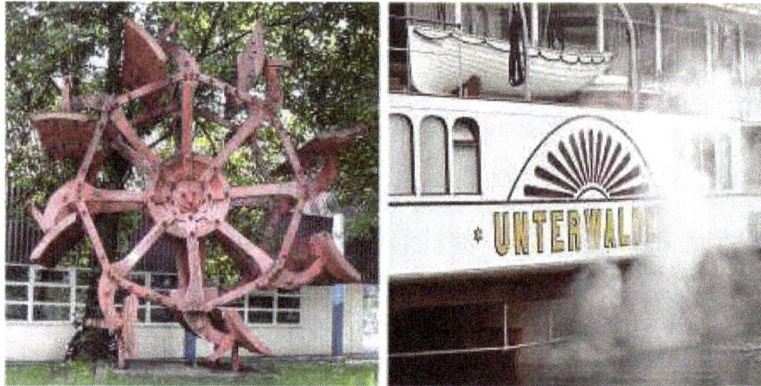

Left: original paddle wheel from a paddle steamer.

Right: detail of a paddle steamer.

The paddle wheel is a large wheel, generally built of a steel framework, upon the outer edge of which are fitted numerous paddle blades (called *floats* or *buckets*). The bottom quarter or so of the wheel travels underwater. Rotation of the paddle wheel produces thrust, forward or backward as required. More advanced paddle wheel designs have featured *feathering* methods that keep each paddle blade oriented closer to vertical while it is in the water; this increases efficiency. The upper part of a paddle wheel is normally enclosed in a paddlebox to minimise splashing.

Paddle wheels have been superseded by screws, which are a much more efficient form of propulsion. Nevertheless, paddle wheels have two advantages over screws, making them suitable for vessels in shallow rivers and constrained waters: first, they are less likely to be clogged by obstacles and debris; and secondly, when contra-rotating, they allow the vessel to spin around its own vertical axis. Some vessels had a single screw in addition to two paddle wheels, to gain the advantages of both types of propulsion.

Sailing

The purpose of sails is to use wind energy to propel the vessel, sled, board, vehicle or rotor.

Water Caterpillar

An early uncommon means of boat propulsion was the water caterpillar. This moved a series of paddles on chains along the bottom of the boat to propel it over the water and preceded the development of tracked vehicles. The first water caterpillar was developed by Desblancs in 1782 and propelled by a steam engine. In the United States the first water caterpillar was patented in 1839 by William Leavenworth of New York.

The water caterpillar boat propulsion system (Popular Science Monthly, December 1918)

Buoyancy

Underwater gliders convert buoyancy to thrust, using wings, or more recently hull shape (SeaExplorer Glider). Buoyancy is made alternatively negative and positive, generating tooth-saw profiles.

Gas Turbine Train

The Turbo Train at Kingston, Ontario, Canada.

A gas turbine train is a passenger train that uses one or more gas turbines as its main source of power. Few passenger trains use this system today, although there has been one recent prototype built by Bombardier Transportation.

Description

A gas turbine train typically consists of two power cars (one at each end of the train), and one or more intermediate passenger cars.

In a gas turbine power car a turbine engine, similar to a turboshaft engine, drives an output shaft that is in turn attached to a hydraulic or electric transmission, or (in the case of the UAC TurboTrain) a mechanical gearbox, which supplies power to drive the wheels.

A gas turbine offers some advantages over a piston engine. There are few moving parts, decreasing the need for lubrication and potentially reducing maintenance costs, and the power-to-weight ratio is much higher. A turbine of a given power output is also physically smaller than an equally powerful piston engine, often allowing the power car to accommodate passengers or cargo as well. However, a turbine's power output and efficiency both drop dramatically with rotational speed, unlike a piston engine, which has a comparatively flat power curve.

Examples

SNCF's turbotrain in Houlgate on the Deauville-Dives railway line in summer 1989.

United Kingdom

British Rail invested in experimentation with the new jet turbines in the early to mid 1950s. Most examples were built for the Western region (because of the oversized loading gauge compared with the rest of the network). The Last attempt before APT-E was GT3 in 1962, but by then British Rail had favoured diesel and electric locomotives, and the project was scrapped.

The gas turbine investigations were rekindled with the construction of British Rail APT-E, prototype of the failed Advanced Passenger Train. Like the French TGV, later models were electric instead. This choice was made because British Leyland, the turbine supplier, ceased production of the model used in the APT-E.

France

SNCF (French National Railways) used a number of gas-turbine trainsets, called the Turbotrain, in non-electrified territory. These typically consisted of a power car at each end with three cars between them. Turbotrain was in use up until 2005. After retirement, 4 sets were sold for further use in Iran.

The first TGV prototype, TGV 001, was also powered by a gas turbine, but steep oil prices prompted the change to overhead electric lines for power delivery.

United States

An RTG Turboliner at Union Station in St. Louis in the 1970s.

In the 1960s United Aircraft built the Turbo passenger train, which was tested by the Pennsylvania Railroad and later used by Amtrak and Via Rail. Via's remained in service into the 1980s and had an excellent maintenance record during this period, but were eventually replaced by the LRC in 1982.

In 1966, the Long Island Rail Road tested an experimental gas turbine railcar *(numbered GT-1)*, powered by two Garrett turbine engines. This car was based on a Budd Pioneer III design, with transmissions similar to Budd's 1950s era RDCs. The car was later modified *(as GT-2)* to add the ability to run on electric third rail as well.

In 1977, the LIRR tested eight more gas turbine-electric/electric dual mode railcars, in an experiment sponsored by the USDOT. Four of these cars had GE-designed powertrains, while the other four had powertrains designed by Garrett *(four more cars had been ordered with GM/ Allison powertrains, but were canceled)*. These cars were similar to LIRR's M1 EMU cars in appearance, with the addition of step wells for loading from low level platforms. The cars suffered from poor fuel economy and mechanical problems, and were withdrawn from service after a short period of time. The four GE-powered cars were converted to M1 EMUs and the Garrett cars were scrapped.

Amtrak purchased two different types of turbine-powered trainsets, which were both called Turboliners. The sets of the first type were similar in appearance to SNCF's T 2000 Turbotrain, though compliance with FRA safety regulations made them heavier and slower than the French trains. None of the first-type Turboliners remain in service. Amtrak also added a number of similarly named Rohr Turboliners (or RTL) to its roster. There were plans to rebuild these as RTL IIIs, but the program has been cancelled and the units are being sold or scrapped.

Canada

Bombardier's experimental JetTrain locomotive toured North America in an early-2000s attempt to raise the technology's public profile.

Canadian National Railways (CN) was one of the operators of the Turbo, which were passed on to Via Rail. They operated on the major Toronto-Montreal route between 1968 and 1982, when they were replaced by the LRC.

In 2002 Bombardier Transportation announced the launch of the JetTrain, a high-speed trainset consisting of tilting carriages and a locomotive powered by a Pratt & Whitney turboshaft engine. While one prototype was built and tested, no JetTrains have yet been sold for actual service.

Gas Turbine-Electric Locomotive

UP 18, preserved at the Illinois Railway Museum.

A gas turbine - electric locomotive, or GTEL, is a locomotive that uses a gas turbine to drive an electric generator or alternator. The electric current thus produced is used to power traction motors. This type of locomotive was first experimented with during the Second World War, but reached its peak in the 1950s to 1960s. Few locomotives use this system today.

Description

A GTEL uses a turbo-electric drivetrain in which a turboshaft engine drives an electrical generator or alternator via a system of gears. The electrical power is distributed to power the traction motors that drive the locomotive. In overall terms the system is very similar to a conventional diesel-electric, with the large diesel engine replaced with a smaller gas turbine of similar power.

A gas turbine offers some advantages over a piston engine. There are few moving parts, decreasing the need for lubrication and potentially reducing maintenance costs, and the power-to-weight ratio is much higher. A turbine of a given power output is also physically smaller than an equally powerful piston engine, allowing a locomotive to be very powerful without being inordinately large. However, a turbine's power output and efficiency both drop dramatically with rotational speed, unlike a piston engine, which has a comparatively flat power curve. This makes GTEL systems useful primarily for long-distance high-speed runs.

Union Pacific operated the largest fleet of such locomotives of any railroad in the world, and was the only railroad to use them for hauling freight. Most other GTELs have been built for small passenger trains, and only a few have seen any real success in that role. With a rise in fuel costs (eventually leading to the 1973 oil crisis), gas turbine locomotives became uneconomical to operate, and many were taken out of service. Additionally, Union Pacific's locomotives required more maintenance than originally anticipated, due to fouling of the turbine blades by the Bunker C oil used as fuel.

Additional problems with gas turbine-electric locomotives included that they were very noisy, and they produced extremely hot exhaust that if the locomotive were parked under an overpass paved with asphalt, it could melt the asphalt.

History

Switzerland

In 1939 the Swiss Federal Railways ordered a GTEL with a 1,620 kW (2,170 hp) of maximum engine power from Brown Boveri. It was completed in 1941, and then underwent testing before entering regular service. The Am 4/6 was the first gas turbine - electric locomotive. It was intended primarily to work light, fast, passenger trains on routes which normally handle insufficient traffic to justify electrification.

United Kingdom

British Rail APT-E

Two gas turbine locomotives of different design, 18000 and 18100 were ordered by the Great Western Railway, but completed for the newly nationalised British Railways.

18000 was built by Brown Boveri and delivered in 1949. It was a 1840 kW (2470 hp) GTEL, ordered by the GWR and used for express passenger services.

18100 was built by Metropolitan-Vickers and delivered in 1951. It had an aircraft-type gas turbine of 2.2 MW (3,000 hp). Maximum speed was 90 miles per hour (140 km/h).

The British Rail APT-E, prototype of the Advanced Passenger Train, was turbine-powered. Like the French TGV, later models were electric instead. This choice was made because British Leyland, the turbine supplier, ceased production of the model used in the APT-E.

France

SNCF In 1952, Renault delivered a prototype four-axle 1150 hp gas-turbine-mechanical locomotive fitted with the Pescara "free turbine" gas- and compressed-air producing system, rather than a co-axial multi-stage compressor integral to the turbine. This model was succeeded by a pair of six-axle 2400 hp locomotives with two turbines and Pescara feeds in 1959. However, these complex locomotives were not a complete success and meanwhile Renault decided to exit the railway equipment business. About a decade later, the first TGV prototype, TGV 001, was powered by a gas turbine, but steep oil prices prompted the change to overhead electric lines for power delivery. However, two large classes of gas-turbine powered intercity railcars were constructed in the early 1970s (ETG and RTG) and were used extensively up to about 2000.

United States

Union Pacific ran a large fleet of turbine-powered freight locomotives starting in the 1950s. These were widely used on long-haul routes, and were cost-effective despite their poor fuel economy due to their use of "leftover" fuels from the petroleum industry. At their height the railroad estimated that they powered about 10% of Union Pacific's freight trains, a much wider use than any other example of this class. As other uses were found for these heavier petroleum byproducts, notably for plastics, the cost of the Bunker C fuel increased until the units became too expensive to operate and they were retired from service by 1969.

In April 1950, Westinghouse completed an experimental 4,000 hp (3,000 kW) turbine locomotive, #4000, known as the Blue Goose, with a B-B-B-B wheel arrangement. The locomotive used two 2,000 hp (1,500 kW) turbine engines, was equipped for passenger train heating with a steam generator that utilized the waste exhaust heat of the right hand turbine, and was geared for 100 miles per hour (160 km/h) While it was demonstrated successfully in both freight and passenger service on the PRR, MKT, and CNW, no production orders followed, and it was scrapped in 1953.

In 1997 the Federal Railroad Administration (FRA) solicited proposals to develop high speed locomotives for routes outside the Northeast Corridor where electrification was not economical. Bombardier Ltd, at the Plattsburg, N.Y. plant where the Acela was produced, developed a prototype (JetTrain) which combined a Pratt & Whitney Canada PW100 gas turbine and a diesel engine with a single gearbox powering four traction motors identical to those in Acela. The diesel provided head end power and low speed traction, with the turbine not being started until after leaving stations. The prototype was completed in June 2000, and safety testing was done at the FRA's Pueblo, CO test track beginning in the summer of 2001. A maximum speed of 156 miles per hour (251 km/h) was reached. The prototype was then taken on a tour of potential sites for high speed service, but no service has yet begun.

Canada

In 2002, Bombardier Transportation announced the launch of the JetTrain, a high-speed trainset consisting of tilting carriages and a locomotive powered by a Pratt & Whitney turboshaft engine. Proposals were made to use the trains for Quebec City-Windsor, Orlando-Miami, and in Alberta, Texas, Nevada and the UK.

However, nothing ever came of any of these proposals, and the JetTrain essentially disappeared, being superseded by the Bombardier Zefiro line of conventionally powered high speed and very high speed trains. The JetTrain no longer appears on any of Bombardier's current web sites or promotional materials, although it can still be found on older web sites bearing the Canadair logos.

Soviet Union

Two gas turbine-electric locomotive types underwent testing in the Soviet Union. The G1-01 freight GTEL was intended to consist of two locomotives of a C-C wheel arrangement, but only one section was built. The test program began in 1959 and lasted into the early 1970s. The GP1 was a similar design, also with a C-C wheel arrangement, introduced in 1964. Two units were built, GP1-0001 and GP1-0002, which were also used in regular service. Both types had a maximum power output of 2,600 kW (3,500 hp).

GT1-001

Russia

In 2006, Russian Railways introduced the GEM-10 switcher GTEL. The turbine's maximum power output is 1,000 kW (1,300 hp) and it runs on liquefied natural gas. The GEM-10 has a C-C wheel arrangement. The TGEM10-0001 is a two-unit (cow-calf) switcher GTEL, with a B-B+B-B wheel arrangement, and uses the same turbine and fuel as the GEM-10.

The GT1-001 freight GTEL, introduced in 2007, runs on liquefied natural gas and has a maximum power output of 8,300 kW (11,100 hp). The locomotive has a B-B-B+B-B-B wheel arrangement, and up to three GT1s can be coupled. On January 23, 2009, the locomotive conducted a test run with a 159 car train weighing 15,000 metric tons (14,800 long tons; 16,500 short tons). Further heavy-haul tests were conducted in December 2010. In a test run conducted in September 2011 the locomotive pulled 170 freight cars weighing 16,000 metric tons (15,700 long tons; 17,600 short tons).

Gas Turbine Locomotive

This section is about gas turbine locomotives with mechanical transmission.

A Gas turbine locomotive is a locomotive powered by a gas turbine. The majority of gas turbine locomotives have had electric transmission but mechanical transmission has also been used, particularly in the early days. The advantage of using gas turbines is that they have very high power-to-bulk and power-to-weight ratios. The disadvantage is that gas turbines generally have lower thermal efficiency than diesel engines, especially when running at less than full load.

Overview

Where electric transmission is used, the engine is usually a single-shaft machine in which one turbine drives both the compressor and the output shaft.

With mechanical transmission, the power turbine must be capable of starting from rest, so a more complex arrangement is necessary. One option is a two-shaft machine, with separate turbines to drive the compressor and the output shaft. Another is to use a separate gas generator, which may be of either rotary or piston type.

Examples

A 44-ton 1-B-1 experimental gas turbine locomotive built in 1952 for testing by the U.S. Army Transportation Corps. Twin 502-2E gas turbines produced 150 hp (110 kW) each. Located at the Museum of Transportation in St. Louis

Examples of gas turbine-mechanical locomotives:

- 1933 Nydqvist and Holm, 1-B-1, Sweden

- 1952 Davenport-Bessler Corp., 1-B-1, United States

- 1951 Renault, France, B-B, 1,000 hp (750 kW)

- 1954 Gotaverken, Sweden, 1-C-1, 1,300 hp (970 kW)

- 1958 Renault, France, C-C, 2,000 hp (1,500 kW)

- 1958 Škoda, C-C, Czechoslovakia, 3,200 hp (2,400 kW)

- 1959 British Rail GT3, 2-C-0, 2,700 hp (2,000 kW)

- 1968 UAC TurboTrain, B'1'1'1'1'1'1'B', 2,000 hp (1,500 kW) (7 car trainset)

Mennons patent

A gas turbine locomotive was patented in 1861 by Marc Antoine Francois Mennons (British patent no. 1633).

The drawings in Mennons' patent show a locomotive of 0-4-2 wheel arrangement with a cylindrical casing resembling a boiler. At the front of the casing is the compressor, which Mennons calls a ventilator. This supplies air to a firebox and the hot gases from the firebox drive a turbine at the back of the casing. The exhaust from the turbine then travels forwards through ducts to preheat the incoming air. The turbine drives the compressor through gearing and an external shaft. There is additional gearing to a jackshaft which drives the wheels through side rods. The fuel is solid (presumably coal, coke or wood) and there is a fuel bunker at the rear.

There is no evidence that the locomotive was actually built but the design includes the essential features of gas turbine locomotives built in the 20th century, including compressor, combustion chamber, turbine and air pre-heater.

The 20Th Century

Work leading to the emergence of the gas turbine locomotive began in France and Sweden in the 1920s but the first locomotive did not appear until 1933. These early experiments used piston engines as gas generators. This idea has not been widely adopted, but it might be worth re-visiting. High fuel consumption was a major factor in the decline of conventional gas-turbine locomotives and the use of a piston engine as a gas generator would probably give better fuel economy than a turbine-type compressor, especially when running at less than full load.

France

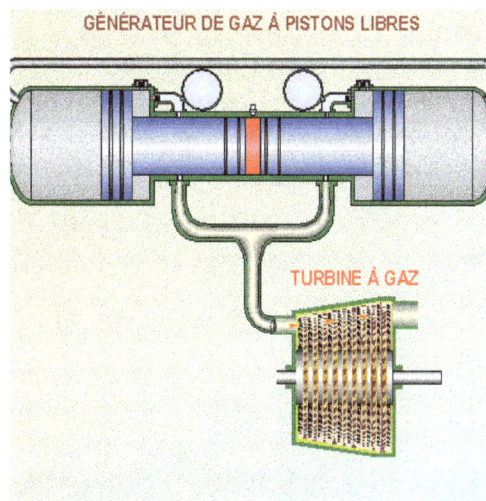

Diagram of a free-piston engine as a gas generator for a gas turbine

The first locomotive, Class 040-GA-1 of 1,000 hp was built by Renault in 1952 and had a Pescara free-piston engine as a gas generator. It was followed by two further locomotives, Class 060-GA-1 of 2,400 hp in 1959-61. Several similar locomotives were built in Russia by Kharkov Locomotive Works.

The Pescara gas generator in 040-GA-1 consisted of a horizontal, single cylinder, two-stroke diesel engine with opposed pistons. It had no crankshaft and the pistons were returned after each power stroke by compression and expansion of air in a separate cylinder. The exhaust from the diesel engine powered the gas-turbine which drove the wheels through a two-speed gearbox and propeller shafts. The free-piston engine was patented in 1934 by Raul Pateras Pescara.

Czechoslovakia

Turbine power was considered for railway traction in the former Czechoslovakia. Two turbine-powered prototypes were built, designated TL 659.001 and .002, featuring C-C wheel arrangement, 3200 hp (2.4 MW) main turbine, helper turbine and Tatra 111 helper diesel engine. The first prototype was finished in February 1958 and was scheduled to be exhibited at Expo '58. This was aborted, because it wasn't ready in time. The first out-of-factory tests were conducted in March 1959 on the Plzeň–Cheb–Sokolov line. On May 15, 1959, the first prototype pulled its heaviest train, 6486 metric tons, but the turbine caught fire only a day later. The engine was never restored and eventually scrapped. The second prototype was built with lessons learned from the first prototype. It left the factory in March 1960 and was the only turbine locomotive to pass the tests for regular service on tracks of the former ČSD. This engine was tried near Kolín and Plzeň with mixed results. This engine was taken out of service in April 1966 and sold to University of Žilina as an educational instrument. The locomotive was scrapped some time later.

Although these experiments had mixed results, they were the most powerful locomotives with purely mechanical powertrain in the world and also the most powerful independent-traction locomotives in Czechoslovakia.

Sweden

The Power gas locomotive was built by Gotaverken. It had a vertical, five cylinder, two-stroke diesel engine with opposed pistons. There was a single crankshaft connected to both upper and lower pistons. The exhaust from the diesel engine powered the gas turbine which drove the wheels through reduction gearing, jack shaft and side rods.

Direct-Drive Gas Turbine

The UAC TurboTrain, built by United Aircraft, entered service with Amtrak and the Canadian National Railway in 1968.

Coal-Firing

In the 1940s and 1950s research was done, in both the USA and UK, aimed at building gas turbine locomotives which could run on pulverized coal. The main problem was to avoid erosion of the turbine blades by particles of ash. Some bench testing was done but the projects were abandoned before any complete locomotives were built. The sources for the following information are Robertson and Sampson.

USA

In the USA, the plan was to use a gas turbine similar to an oil-fuelled one and to remove ash particles with filters. Details of the US research (done in 1946) were passed to Britain's London, Midland and Scottish Railway.

UK

On 23 December 1952 the UK Ministry of Fuel and Power placed an order for a coal-fired gas turbine locomotive to be used on British Railways. The locomotive was to be built by the North British Locomotive Company and the turbine would be supplied by C. A. Parsons and Company.

According to Sampson, the plan was to use indirect heating. The pulverized coal would be burned in a combustion chamber and the hot gases passed to a heat exchanger. Here, the heat would be transferred to a separate body of compressed air which would power the turbine. Essentially, it would have been a hot air engine using a turbine instead of a piston.

Robertson shows a diagram which confirms Sampson's information but also refers to problems with erosion of turbine blades by ash. This is strange because, with a conventional shell and tube heat exchanger, there would be no risk of ash entering the turbine circuit.

Working Cycle

There were two separate, but linked, circuits - the combustion circuit and the turbine circuit.

1. Combustion circuit. Pulverized coal and air were mixed and burned in a combustion chamber and the hot gases passed to a heat exchanger where heat was transferred to the compressed air in the turbine circuit. After leaving the heat exchanger the combustion gases entered a boiler to generate steam for train heating.

2. Turbine circuit. Air entered the compressor and was compressed. The compressed air passed to the heat exchanger where it was heated by the combustion gases. The heated compressed air drove two turbines - one to drive the compressor and the other to power the locomotive. The turbine exhaust (which was hot air) then entered the combustion chamber to support the combustion.

Specification

The locomotive was never built but the specification was as follows:

- Wheel arrangement: C-C, later changed to 1A1A-A1A1
- Horsepower: 1,800, later reduced to 1,500
- Weight: 117 tons, later increased to 150 tons

The projected output was:

- Tractive effort,
    - 30,000 lbf (130 kN) at 72 mph (116 km/h)
    - 45,000 lbf (200 kN) at 50 mph (80 km/h)
- Thermal efficiency,
    - 10% at 1/10 load
    - 16% at half load
    - 19% at full load

The transmission was to be mechanical, via a two-speed gearbox, giving a high speed for passenger working and a lower speed for freight. The tractive effort figures, quoted above, look suspiciously high for the specified speeds. It seems more likely that the figures quoted are for starting tractive effort and maximum speed in high gear and low gear respectively.

There is a model of the proposed locomotive at Glasgow Museum of Transport and some records are held at the National Railway Museum.

The British Rail GT3 was a much simpler machine consisting essentially of a standard oil-fired gas turbine mounted on a standard steam locomotive chassis, built as a demonstrator by English Electric in 1961. Its almost crude simplicity enabled it to avoid much of the unreliability which had plagued the complex experimental GTELs 18000 and 18100 in earlier years, but it nevertheless failed to be competitive against conventional traction and was scrapped.

Aircraft

NASA test aircraft

An aircraft is a machine that is able to fly by gaining support from the air. It counters the force of gravity by using either static lift or by using the dynamic lift of an airfoil, or in a few cases the downward thrust from jet engines.

The Mil Mi-8 is the most-produced helicopter in history

Voodoo, a modified P 51 Mustang is the 2014 Reno Air Race Champion

The human activity that surrounds aircraft is called *aviation*. Crewed aircraft are flown by an onboard pilot, but unmanned aerial vehicles may be remotely controlled or self-controlled by onboard computers. Aircraft may be classified by different criteria, such as lift type, aircraft propulsion, usage and others.

History

Flying model craft and stories of manned flight go back many centuries, however the first manned ascent – and safe descent – in modern times took place by larger hot-air balloons developed in the 18th century. Each of the two World Wars led to great technical advances. Consequently, the history of aircraft can be divided into five eras:

- Pioneers of flight, from the earliest experiments to 1914.

- First World War, 1914 to 1918.

- Aviation between the World Wars, 1918 to 1939.

- Second World War, 1939 to 1945.

- Postwar era, also called the jet age, 1945 to the present day.

Methods of Lift

Lighter than air – aerostats

A hot air balloon in flight

Aerostats use buoyancy to float in the air in much the same way that ships float on the water. They are characterized by one or more large gasbags or canopies, filled with a relatively low-density gas such as helium, hydrogen, or hot air, which is less dense than the surrounding air. When the weight of this is added to the weight of the aircraft structure, it adds up to the same weight as the air that the craft displaces.

Small hot-air balloons called sky lanterns were first invented in ancient China prior to the 3rd century BC and used primarily in cultural celebrations, and were only the second type of aircraft to fly, the first being kites which were first invented in ancient China over two thousand years ago.

A balloon was originally any aerostat, while the term airship was used for large, powered aircraft designs – usually fixed-wing. In 1919 Frederick Handley Page was reported as referring to "ships of the air," with smaller passenger types as "Air yachts." In the 1930s, large intercontinental flying boats were also sometimes referred to as "ships of the air" or "flying-ships". – though none had yet been built. The advent of powered balloons, called dirigible balloons, and later of rigid hulls allowing a great increase in size, began to change the way these words were used. Huge powered aerostats, characterized by a rigid outer framework and separate aerodynamic skin surrounding the gas bags, were produced, the Zeppelins being the largest and most famous. There were still no fixed-wing aircraft or non-rigid balloons large enough to be called airships, so "airship" came to be synonymous with these aircraft. Then several accidents, such as the Hindenburg disaster in 1937, led to the demise of these airships. Nowadays a "balloon" is an unpowered aerostat and an "airship" is a powered one.

A powered, steerable aerostat is called a *dirigible*. Sometimes this term is applied only to non-rigid balloons, and sometimes *dirigible balloon* is regarded as the definition of an airship (which may then be rigid or non-rigid). Non-rigid dirigibles are characterized by a moderately aerodynamic gasbag with stabilizing fins at the back. These soon became known as *blimps*. During the Second World War, this shape was widely adopted for tethered balloons; in windy weather, this both reduces the strain on the tether and stabilizes the balloon. The nickname *blimp* was adopted along with the shape. In modern times, any small dirigible or airship is called a blimp, though a blimp may be unpowered as well as powered.

Heavier-Than-Air – Aerodynes

Heavier-than-air aircraft, such as airplanes, must find some way to push air or gas downwards, so that a reaction occurs (by Newton's laws of motion) to push the aircraft upwards. This dynamic movement through the air is the origin of the term *aerodyne*. There are two ways to produce dynamic upthrust: aerodynamic lift, and powered lift in the form of engine thrust.

Aerodynamic lift involving wings is the most common, with fixed-wing aircraft being kept in the air by the forward movement of wings, and rotorcraft by spinning wing-shaped rotors sometimes called rotary wings. A wing is a flat, horizontal surface, usually shaped in cross-section as an aerofoil. To fly, air must flow over the wing and generate lift. A *flexible wing* is a wing made of fabric or thin sheet material, often stretched over a rigid frame. A *kite* is tethered to the ground and relies on the speed of the wind over its wings, which may be flexible or rigid, fixed, or rotary.

With powered lift, the aircraft directs its engine thrust vertically downward. V/STOL aircraft, such as the Harrier Jump Jet and F-35B take off and land vertically using powered lift and transfer to aerodynamic lift in steady flight.

A pure rocket is not usually regarded as an aerodyne, because it does not depend on the air for its lift (and can even fly into space); however, many aerodynamic lift vehicles have been powered or assisted by rocket motors. Rocket-powered missiles that obtain aerodynamic lift at very high speed due to airflow over their bodies are a marginal case.

Fixed-Wing

An Airbus A380, the world's largest passenger airliner, flown at the 2007 Air Expo.

The forerunner of the fixed-wing aircraft is the kite. Whereas a fixed-wing aircraft relies on its forward speed to create airflow over the wings, a kite is tethered to the ground and relies on the wind blowing over its wings to provide lift. Kites were the first kind of aircraft to fly, and were invented in China around 500 BC. Much aerodynamic research was done with kites before test aircraft, wind tunnels, and computer modelling programs became available.

The first heavier-than-air craft capable of controlled free-flight were gliders. A glider designed by Cayley carried out the first true manned, controlled flight in 1853.

Practical, powered, fixed-wing aircraft (the aeroplane or airplane) were invented by Wilbur and Orville Wright. Besides the method of propulsion, fixed-wing aircraft are in general characterized by their wing configuration. The most important wing characteristics are:

- Number of wings – Monoplane, biplane, etc.

- Wing support – Braced or cantilever, rigid, or flexible.

- Wing planform – including aspect ratio, angle of sweep, and any variations along the span (including the important class of delta wings).

- Location of the horizontal stabilizer, if any.

- Dihedral angle – positive, zero, or negative (anhedral).

A variable geometry aircraft can change its wing configuration during flight.

A *flying wing* has no fuselage, though it may have small blisters or pods. The opposite of this is a *lifting body*, which has no wings, though it may have small stabilizing and control surfaces.

Wing-in-ground-effect vehicles are not considered aircraft. They "fly" efficiently close to the surface of the ground or water, like conventional aircraft during takeoff. An example is the Russian ekranoplan (nicknamed the "Caspian Sea Monster"). Man-powered aircraft also rely on ground effect to remain airborne with a minimal pilot power, but this is only because they are so under-powered — in fact, the airframe is capable of flying higher.

Rotorcraft

An autogyro

Rotorcraft, or rotary-wing aircraft, use a spinning rotor with aerofoil section blades (a *rotary wing*) to provide lift. Types include helicopters, autogyros, and various hybrids such as gyrodynes and compound rotorcraft.

Helicopters have a rotor turned by an engine-driven shaft. The rotor pushes air downward to create lift. By tilting the rotor forward, the downward flow is tilted backward, producing thrust for forward flight. Some helicopters have more than one rotor and a few have rotors turned by gas jets at the tips.

Autogyros have unpowered rotors, with a separate power plant to provide thrust. The rotor is tilted backward. As the autogyro moves forward, air blows upward across the rotor, making it spin. This spinning increases the speed of airflow over the rotor, to provide lift. Rotor kites are unpowered autogyros, which are towed to give them forward speed or tethered to a static anchor in high-wind for kited flight.

Cyclogyros rotate their wings about a horizontal axis.

Compound rotorcraft have wings that provide some or all of the lift in forward flight. They are nowadays classified as *powered lift* types and not as rotorcraft. *Tiltrotor* aircraft (such as the V-22 Osprey), tiltwing, tailsitter, and coleopter aircraft have their rotors/propellers horizontal for vertical flight and vertical for forward flight.

Other Methods of Lift

X-24B lifting body, specialized glider

- A *lifting body* is an aircraft body shaped to produce lift. If there are any wings, they are too small to provide significant lift and are used only for stability and control. Lifting bodies are not efficient: they suffer from high drag, and must also travel at high speed to generate enough lift to fly. Many of the research prototypes, such as the Martin-Marietta X-24, which led up to the Space Shuttle, were lifting bodies (though the shuttle itself is not), and some supersonic missiles obtain lift from the airflow over a tubular body.

- *Powered lift* types rely on engine-derived lift for vertical takeoff and landing (VTOL). Most types transition to fixed-wing lift for horizontal flight. Classes of powered lift types include VTOL jet aircraft (such as the Harrier jump-jet) and tiltrotors (such as the V-22 Osprey), among others. A few experimental designs rely entirely on engine thrust to provide lift throughout the whole flight, including personal fan-lift hover platforms and jetpacks. VTOL research designs include the flying Bedstead.

- The *Flettner airplane* uses a rotating cylinder in place of a fixed wing, obtaining lift from the magnus effect.

- The *ornithopter* obtains thrust by flapping its wings.

Propulsion

Unpowered Aircraft

Gliders are heavier-than-air aircraft that do not employ propulsion once airborne. Take-off may be by launching forward and downward from a high location, or by pulling into the air on a tow-line, either by a ground-based winch or vehicle, or by a powered "tug" aircraft. For a glider to maintain its forward air speed and lift, it must descend in relation to the air (but not necessarily in relation to the ground). Many gliders can 'soar' – gain height from updrafts such as thermal currents. The first practical, controllable example was designed and built by the British scientist and pioneer George Cayley, whom many recognise as the first aeronautical engineer. Common examples of gliders are sailplanes, hang gliders and paragliders.

Balloons drift with the wind, though normally the pilot can control the altitude, either by heating the air or by releasing ballast, giving some directional control (since the wind direction changes with altitude). A wing-shaped hybrid balloon can glide directionally when rising or falling; but a spherically shaped balloon does not have such directional control.

Kites are aircraft that are tethered to the ground or other object (fixed or mobile) that maintains tension in the tether or kite line; they rely on virtual or real wind blowing over and under them to generate lift and drag. Kytoons are balloon-kite hybrids that are shaped and tethered to obtain kiting deflections, and can be lighter-than-air, neutrally buoyant, or heavier-than-air.

Powered Aircraft

Powered aircraft have one or more onboard sources of mechanical power, typically aircraft engines although rubber and manpower have also been used. Most aircraft engines are either lightweight piston engines or gas turbines. Engine fuel is stored in tanks, usually in the wings but larger aircraft also have additional fuel tanks in the fuselage.

Propeller Aircraft

A turboprop-engined DeHavilland Twin Otter adapted as a floatplane

Propeller aircraft use one or more propellers (airscrews) to create thrust in a forward direction. The propeller is usually mounted in front of the power source in *tractor configuration* but can be mounted behind in *pusher configuration*. Variations of propeller layout include *contra-rotating propellers* and *ducted fans*.

Many kinds of power plant have been used to drive propellers. Early airships used man power or steam engines. The more practical internal combustion piston engine was used for virtually all fixed-wing aircraft until World War II and is still used in many smaller aircraft. Some types use turbine engines to drive a propeller in the form of a turboprop or propfan. Human-powered flight has been achieved, but has not become a practical means of transport. Unmanned aircraft and models have also used power sources such as electric motors and rubber bands.

Jet Aircraft

Lockheed Martin F-22A Raptor

Jet aircraft use airbreathing jet engines, which take in air, burn fuel with it in a combustion chamber, and accelerate the exhaust rearwards to provide thrust.

Turbojet and turbofan engines use a spinning turbine to drive one or more fans, which provide additional thrust. An afterburner may be used to inject extra fuel into the hot exhaust, especially on military "fast jets". Use of a turbine is not absolutely necessary: other designs include the pulse jet and ramjet. These mechanically simple designs cannot work when stationary, so the aircraft must be launched to flying speed by some other method. Other variants have also been used, including the motorjet and hybrids such as the Pratt & Whitney J58, which can convert between turbojet and ramjet operation.

Compared to propellers, jet engines can provide much higher thrust, higher speeds and, above about 40,000 ft (12,000 m), greater efficiency. They are also much more fuel-efficient than rockets. As a consequence nearly all large, high-speed or high-altitude aircraft use jet engines.

Rotorcraft

Some rotorcraft, such as helicopters, have a powered rotary wing or *rotor*, where the rotor disc can be angled slightly forward so that a proportion of its lift is directed forwards. The rotor may, like a propeller, be powered by a variety of methods such as a piston engine or turbine. Experiments have also used jet nozzles at the rotor blade tips.

Other Types of Powered Aircraft

Rocket-powered aircraft have occasionally been experimented with, and the Messerschmitt *Komet* fighter even saw action in the Second World War. Since then, they have been restricted to research aircraft, such as the North American X-15, which traveled up into space where air-breathing engines cannot work (rockets carry their own oxidant). Rockets have more often been used as a supplement to the main power plant, typically for the rocket-assisted take off of heavily loaded aircraft, but also to provide high-speed dash capability in some hybrid designs such as the Saunders-Roe SR.53.

The *ornithopter* obtains thrust by flapping its wings. It has found practical use in a model hawk used to freeze prey animals into stillness so that they can be captured, and in toy birds.

Design and Construction

Aircraft are designed according to many factors such as customer and manufacturer demand, safety protocols and physical and economic constraints. For many types of aircraft the design process is regulated by national airworthiness authorities.

The key parts of an aircraft are generally divided into three categories:

- The *structure* comprises the main load-bearing elements and associated equipment.

- The *propulsion system* (if it is powered) comprises the power source and associated equipment, as described above.

- The *avionics* comprise the control, navigation and communication systems, usually electrical in nature.

Structure

The approach to structural design varies widely between different types of aircraft. Some, such as paragliders, comprise only flexible materials that act in tension and rely on aerodynamic pressure to hold their shape. A balloon similarly relies on internal gas pressure but may have a rigid basket or gondola slung below it to carry its payload. Early aircraft, including airships, often employed flexible doped aircraft fabric covering to give a reasonably smooth aeroshell stretched over a rigid frame. Later aircraft employed semi-monocoque techniques, where the skin of the aircraft is stiff enough to share much of the flight loads. In a true monocoque design there is no internal structure left.

The key structural parts of an aircraft depend on what type it is.

Aerostats

Lighter-than-air types are characterised by one or more gasbags, typically with a supporting structure of flexible cables or a rigid framework called its hull. Other elements such as engines or a gondola may also be attached to the supporting structure.

Aerodynes

Heavier-than-air types are characterised by one or more wings and a central fuselage. The fuselage typically also carries a tail or empennage for stability and control, and an undercarriage for takeoff and landing. Engines may be located on the fuselage or wings. On a fixed-wing aircraft the wings are rigidly attached to the fuselage, while on a rotorcraft the wings are attached to a rotating vertical shaft. Smaller designs sometimes use flexible materials for part or all of the structure, held in place either by a rigid frame or by air pressure. The fixed parts of the structure comprise the airframe.

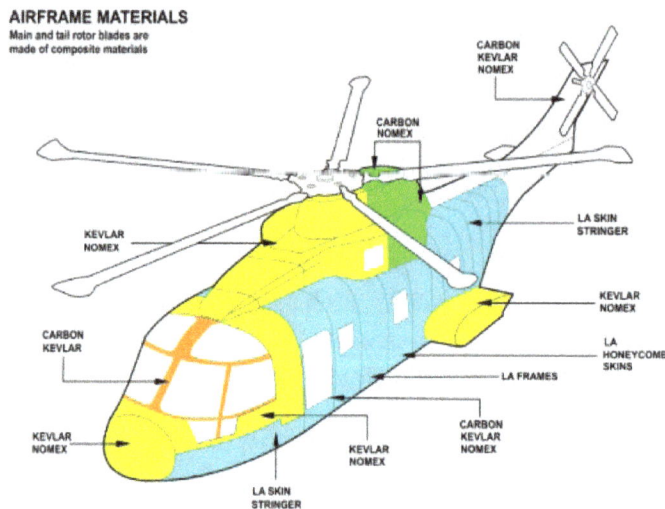

Airframe diagram for an AgustaWestland AW101 helicopter

Avionics

The avionics comprise the flight control systems and related equipment, including the cockpit instrumentation, navigation, radar, monitoring, and communication systems.

Flight Characteristics

Flight Envelope

The flight envelope of an aircraft refers to its capabilities in terms of airspeed and load factor or altitude. The term can also refer to other measurements such as maneuverability. When a craft is pushed, for instance by diving it at high speeds, it is said to be flown *outside the envelope*, something considered unsafe.

Range

The Boeing 777-200LR is the longest-range airliner, capable of flights of more than halfway around the world.

The range is the distance an aircraft can fly between takeoff and landing, as limited by the time it can remain airborne.

For a powered aircraft the time limit is determined by the fuel load and rate of consumption.

For an unpowered aircraft, the maximum flight time is limited by factors such as weather conditions and pilot endurance. Many aircraft types are restricted to daylight hours, while balloons are limited by their supply of lifting gas. The range can be seen as the average ground speed multiplied by the maximum time in the air.

Flight Dynamics

Flight dynamics is the science of air vehicle orientation and control in three dimensions. The three critical flight dynamics parameters are the angles of rotation around three axes about the vehicle's center of mass, known as *pitch*, *roll*, and *yaw* (quite different from their use as Tait-Bryan angles).

- Roll is a rotation about the longitudinal axis (equivalent to the rolling or heeling of a ship) giving an up-down movement of the wing tips measured by the roll or bank angle.

- Pitch is a rotation about the sideways horizontal axis giving an up-down movement of the aircraft nose measured by the angle of attack.

- Yaw is a rotation about the vertical axis giving a side-to-side movement of the nose known as sideslip.

Flight dynamics is concerned with the stability and control of an aircraft's rotation about each of these axes.

Stability

An aircraft that is unstable tends to diverge from its current flight path and so is difficult to fly. A very stable aircraft tends to stay on its current flight path and is difficult to manoeuvre—so it is important for any design to achieve the desired degree of stability. Since the widespread use of digital computers, it is increasingly common for designs to be inherently unstable and rely on computerised control systems to provide artificial stability.

The tail assembly of a Boeing 747–200

A fixed wing is typically unstable in pitch, roll, and yaw. Pitch and yaw stabilities of conventional fixed wing designs require horizontal and vertical stabilisers, which act similarly to the feathers on an arrow. These stabilizing surfaces allow equilibrium of aerodynamic forces and to stabilise the flight dynamics of pitch and yaw. They are usually mounted on the tail section (empennage), although in the canard layout, the main aft wing replaces the canard foreplane as pitch stabilizer. Tandem wing and Tailless aircraft rely on the same general rule to achieve stability, the aft surface being the stabilising one.

A rotary wing is typically unstable in yaw, requiring a vertical stabiliser.

A balloon is typically very stable in pitch and roll due to the way the payload is hung underneath.

Control

Flight control surfaces enable the pilot to control an aircraft's flight attitude and are usually part of the wing or mounted on, or integral with, the associated stabilizing surface. Their development was a critical advance in the history of aircraft, which had until that point been uncontrollable in flight.

Aerospace engineers develop control systems for a vehicle's orientation (attitude) about its center of mass. The control systems include actuators, which exert forces in various directions, and generate rotational forces or moments about the aerodynamic center of the aircraft, and thus rotate the aircraft in pitch, roll, or yaw. For example, a pitching moment is a vertical force applied at a distance forward or aft from the aerodynamic center of the aircraft, causing the aircraft to pitch up or down. Control systems are also sometimes used to increase or decrease drag, for example to slow the aircraft to a safe speed for landing.

The two main aerodynamic forces acting on any aircraft are lift supporting it in the air and drag opposing its motion. Control surfaces or other techniques may also be used to affect these forces directly, without inducing any rotation.

Impacts of Aircraft Use

Aircraft permit long distance, high speed travel and may be a more fuel efficient mode of transportation in some circumstances. Aircraft have environmental and climate impacts beyond fuel efficiency considerations, however. They are also relatively noisy compared to other forms of travel and high altitude aircraft generate contrails, which experimental evidence suggests may alter weather patterns.

Uses for Aircraft

Aircraft are produced in several different types optimized for various uses; military aircraft, which includes not just combat types but many types of supporting aircraft, and civil aircraft, which include all non-military types, experimental and model.

Military

Boeing B-17E in flight

A military aircraft is any aircraft that is operated by a legal or insurrectionary armed service of any type. Military aircraft can be either combat or non-combat:

Combat aircraft are aircraft designed to destroy enemy equipment using its own armament. Combat aircraft divide broadly into fighters and bombers, with several in-between types such as fighter-bombers and ground-attack aircraft (including attack helicopters).

Non-combat aircraft are not designed for combat as their primary function, but may carry weapons for self-defense. Non-combat roles include search and rescue, reconnaissance, observation, transport, training, and aerial refueling. These aircraft are often variants of civil aircraft.

Most military aircraft are powered heavier-than-air types. Other types such as gliders and balloons have also been used as military aircraft; for example, balloons were used for observation during the American Civil War and World War I, and military gliders were used during World War II to land troops.

Civil

Agusta A109 helicopter of the Swiss air rescue service

Civil aircraft divide into *commercial* and *general* types, however there are some overlaps.

Commercial aircraft include types designed for scheduled and charter airline flights, carrying passengers, mail and other cargo. The larger passenger-carrying types are the airliners, the largest of which are wide-body aircraft. Some of the smaller types are also used in general aviation, and some of the larger types are used as VIP aircraft.

General aviation is a catch-all covering other kinds of private (where the pilot is not paid for time or expenses) and commercial use, and involving a wide range of aircraft types such as business jets (bizjets), trainers, homebuilt, gliders, warbirds and hot air balloons to name a few. The vast majority of aircraft today are general aviation types.

Experimental

An experimental aircraft is one that has not been fully proven in flight, or that carries an FAA airworthiness certificate in the "Experimental" category. Often, this implies that the aircraft is testing new aerospace technologies, though the term also refers to amateur and kit-built aircraft—many based on proven designs.

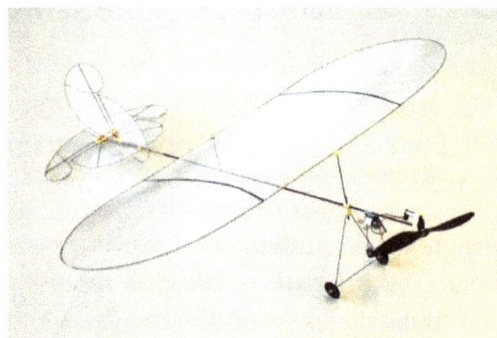

A model aircraft, weighing six grams

Model

A model aircraft is a small unmanned type made to fly for fun, for static display, for aerodynamic research or for other purposes. A scale model is a replica of some larger design.

Propfan

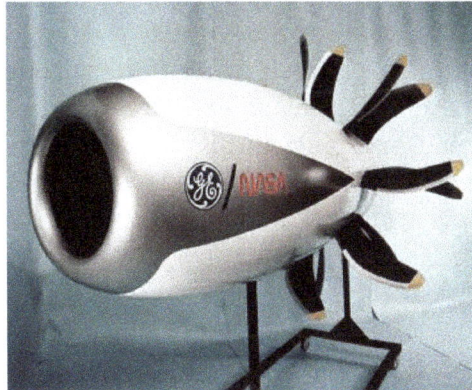

NASA / GE unducted fan

A propfan or open rotor engine is a type of aircraft engine related in concept to both the turboprop and turbofan, but distinct from both. The European Aviation Safety Agency (EASA) defines it as *"A turbine engine featuring contra rotating fan stages not enclosed within a casing."* The engine uses a gas turbine to drive an unshrouded (open) contra-rotating propeller like a turboprop, but the design of the propeller itself is more tightly coupled to the turbine design and the two are certified as a single unit.

A propfan is typically designed with a large number of short, highly twisted blades, similar to a turbofan's bypass compressor (the "fan" itself). For this reason, the propfan has been variously described as an "unducted fan" or an "ultra-high-bypass (UHB) turbofan". In technical papers it is described as "a small diameter, highly loaded multiple bladed variable pitch propulsor having swept blades with thin advanced airfoil sections, integrated with a nacelle contoured to retard the airflow through the blades thereby reducing compressibility losses and designed to operate with a turbine engine and using a single stage reduction gear resulting in high performance." The design is intended to offer the speed and performance of a turbofan, with the fuel economy of a turboprop. The propfan concept was first revealed by Carl Rohrbach and Bruce Metzger of the Hamilton Standard Division of United Technologies in 1975 and was patented by Robert Cornell and Carl Rohrbach of Hamilton Standard in 1979. Later work by General Electric on similar propulsors was done under the name unducted fan, which was a modified turbofan engine, with the fan placed outside the engine nacelle on the same axis as the compressor blades.

Limitations and Solutions

Propeller Blade Tip Speed Limit

Turboprops have an optimum speed below about 450 mph (700 km/h). The reason is that all propellers lose efficiency at high speed, due to an effect known as wave drag that occurs just below supersonic speeds. This powerful form of drag has a sudden onset, and led to the concept of a sound barrier when it was first encountered in the 1940s. In the case of a propeller, this effect can happen any time the propeller is spun fast enough that the blade tips near the speed of sound, even if the aircraft is motionless on the ground.

The most effective way to counteract this problem (to some degree) is by adding more blades to the propeller, allowing it to deliver more power at a lower rotational speed. This is why many World War II fighter designs started with two or three-blade propellers and by the end of the war were using up to five blades in some cases as the engines were upgraded and new propellers were needed to more efficiently convert that power. The major downside to this approach is that adding blades makes the propeller harder to balance and maintain and the additional blades cause minor performance penalties (due to drag and efficiency issues). But even with these sorts of measures at some point the forward speed of the plane combined with the rotational speed of the propeller will once again result in wave drag problems. For most aircraft this will occur at speeds over about 450 mph (700 km/h).

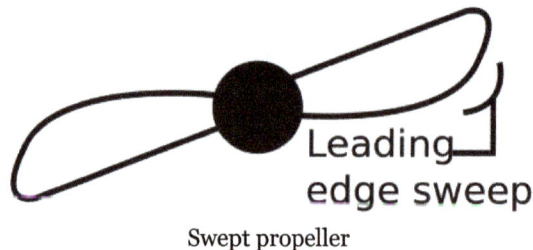

Swept propeller

A method of decreasing wave drag was discovered by German researchers in 1935—sweeping the wing backwards. Today, almost all aircraft designed to fly much above 450 mph (700 km/h) use a swept wing. In the 1970s, Hamilton Standard started researching propellers with similar sweep. Since the inside of the propeller is moving slower than the outside, the blade is progressively more swept toward the outside, leading to a curved shape similar to a scimitar - a practice that was first used as far back as 1909, in the Chauvière make of two-bladed wood propeller used on the Blériot XI.

Jet Aircraft Fuel Economy

Jet aircraft fly faster than conventional propeller-driven aircraft. However, they use more fuel, so that for the same fuel consumption a propeller installation produces more thrust. As fuel costs become an increasingly important aspect of commercial aviation, engine designers continue to seek ways to improve aero-engine efficiency.

The propfan concept was developed to deliver 35% better fuel efficiency than contemporary turbofans. In static and air tests on a modified Douglas DC-9, propfans reached a 30% improvement over the OEM turbofans. This efficiency came at a price, as one of the major problems with the propfan is noise, particularly in an era where aircraft are required to comply with increasingly strict aircraft noise regulations. However, in 2012 GE expects that propfans can meet these noise levels by 2030 when new narrowbody generations from Boeing and Airbus become available. Airlines consistently ask for low noise, and then maximum fuel efficiency.

The Hamilton Standard Division of United Technologies developed the propfan concept in the early 1970s. Numerous design variations of the propfan were tested by Hamilton Standard, in conjunction with NASA in this decade. This testing led to the Propfan Test Assessment (PTA) program, where Lockheed-Georgia proposed modifying a Gulfstream II to act as in-flight testbed for the propfan concept and McDonnell Douglas proposed modifying a DC-9 for the same purpose. NASA chose the Lockheed proposal, where the aircraft had a nacelle added to the left wing, containing a 6000 hp Allison 570 turboprop engine (derived from the XT701 turboshaft developed for the Boeing Vertol XCH-62 program), powering a 9-foot diameter Hamilton Standard SR-7 propfan. The aircraft, so configured, first flew in March 1987. After an extensive test program, the modifications were removed from the aircraft.

General Electric's GE36 Unducted Fan was a variation on the original propfan concept, and appears similar to a pusher configuration piston engine. GE's UDF had a novel direct drive arrangement, where the reduction gearbox was replaced by a low-speed seven-stage free turbine. One set of turbine rotors drove the forward set of propellers, while the rear set was driven by the other set of rotors which rotated in the opposite direction. The turbine had 14 blade rows with 7 stages. Each stage was a pair of counter-rotating rows. Boeing intended to offer GE's pusher UDF engine on the 7J7 platform, and McDonnell Douglas was going to do likewise on their MD-94X airliner. The GE36 was first flight tested mounted on the #3 engine station of a Boeing 727-100 in 1986.

McDonnell Douglas developed a proof-of-concept aircraft by modifying its company-owned MD-80. They removed the JT8D turbofan engine from the left side of the fuselage and replaced it with the GE36. A number of test flights were conducted, initially out of Mojave, California, which proved the airworthiness, aerodynamic characteristics, and noise signature of the design. Following the initial tests, a first-class cabin was installed inside the aft fuselage and airline executives were offered the opportunity to experience the UDF-powered aircraft first-hand. The test and marketing flights of the GE-outfitted demonstrator aircraft concluded in 1988, exhibiting a 30% reduction in fuel consumption over turbo-fan powered MD-80, full Stage III noise compliance, and low levels of interior noise/vibration. Due to jet fuel price drops and shifting marketing priorities, Douglas shelved the program the following year.

In the 1980s, Allison collaborated with Pratt & Whitney on demonstrating the 578-DX propfan. Unlike the competing GE36 UDF, the 578-DX was fairly conventional, having a reduction gearbox between the LP turbine and the propfan blades. The 578-DX was successfully flight tested on a McDonnell Douglas MD-80. However, none of the above projects came to fruition, mainly because of excessive cabin noise (compared to turbofans) and low fuel prices.

The Progress D-27 propfan, developed in the U.S.S.R., was designed with the propfan blades at the front of the engine in a tractor configuration. Two rear-mounted D-27 propfans propelled the Ukrainian Antonov An-180, which was scheduled for a 1995 entry into service. Another propfan application was the Russian Yakovlev Yak-46. During the 1990s, Antonov also developed the An-70, powered by four Progress D-27s in a tractor configuration; the Russian Air Force placed an order for 164 aircraft in 2003, which was subsequently canceled. However, the An-70 remains available for further investment and production.

With increasing prices for jet fuel and the emphasis on engine/airframe efficiency to reduce emissions, there is renewed interest in the propfan concept for jetliners that might come into service beyond the Boeing 787 and Airbus A350XWB. For instance, Airbus has patented aircraft designs with twin rear-mounted counter-rotating propfans.

Progress D27 Propfans fitted to an Antonov An-70

References

- Robertson, K. The Great Western Railway Gas Turbines, published by Alan Sutton, 1989, ISBN 0-86299-541-8

- Crane, Dale: Dictionary of Aeronautical Terms, third edition, page 194. Aviation Supplies & Academics, 1997. ISBN 1-56027-287-2

- Aviation Publishers Co. Limited, From the Ground Up, page 10 (27th revised edition) ISBN 0-9690054-9-0

- "First Improved Oyashio-class boat takes to the water". IHS. June 12, 2007. Archived from the original on 7 June 2011. Retrieved June 3, 2011.

- "Experimental gas turbine locomotive undertakes haulage tests". Railway Gazette International. 2009-01-14. Retrieved 2011-02-04.

Permissions

Index

www.ingramcontent.com/pod-product-compliance
Lightning Source LLC
Chambersburg PA
CBHW061305190326
41458CB00011B/3768